AN INTRODUCTION TO
FIBER OPTICS SYSTEM DESIGN

AN INTRODUCTION TO FIBER OPTICS SYSTEM DESIGN

Bruce E. BRILEY
AT&T Bell Laboratories
Naperville, Illinois,
U.S.A.

1988

NORTH-HOLLAND
AMSTERDAM · NEW YORK · OXFORD · TOKYO

ISBN: 0 444 70498 1

Publishers:

ELSEVIER SCIENCE PUBLISHERS B.V.
P.O. Box 1991
1000 BZ Amsterdam
The Netherlands

Sole distributors for the U.S.A. and Canada:

ELSEVIER SCIENCE PUBLISHING COMPANY, INC.
52 Vanderbilt Avenue
New York, N.Y. 10017
U.S.A.

PRINTED IN THE NETHERLANDS

TO MARILYN, DENISE, SCOTT, STEVEN, AND TODD

PREFACE

It is a privilege to have been technically active during two technology revolutions brought on, respectively, by the advent of the transistor and of practical fiber optics. The effects of the transistor are so pervasive as to intrude into almost every phase of personal as well as professional life: communications and entertainment equipment, computers, the automobile, etc. The effects of fiber optics are less pervasive as yet, but they may well eventually rival those of the transistor.

Cheap, high-quality (low-noise) bandwidth will likely be available to all eventually via fiber optics. The uses of such bandwidth (transmission of voice, data and video) that immediately suggest themselves may well be only the most elementary applications when viewed against the backdrop of the dimly perceived future.

It will likely take a generation or two before the "bandwidth is expensive" mindset is overcome, and heroic efforts to conserve it cease. Of course, if history can teach by analogy, it should be noted that any worthwhile commodity provided in apparent excess is rapidly depleted (e.g., computer real time).

The author began teaching a graduate-level course in fiber optics at a time when there were no usable textbooks on the subject; the notes generated eventually crystallized into this book. The target of this text is the student and professional electrical/electronic engineer. It therefore attempts to build upon an assumed model of a typical student or practicing engineer who has been exposed to conventional transmission line, waveguide, circuit, and semiconductor principles and can therefore readily grasp fiber-optics system fundamentals on a comparative-anatomy basis.

The text was produced using electronic typesetting, allowing the final version to be up to date and in print within three months (this technique may soon become universal).

As something of an experiment, solutions (not just answers) to the exercises are included as an appendix, on the theory that the better students will attempt a problem before looking toward the appendix (the best approach), the struggling student will not be left out in the cold, and the tutorial thinking behind the given solution (or at least *a* solution) may serve a useful purpose.

The effects of radiation upon fiber optics components and systems (a topic often neglected) are discussed in association with each pertinent section.

I wish to express my thanks to Harvey Lehman, Avi Vaidya, Dave Vlack, Bob Staehler, and John Degan, who supported this effort unstintingly, to the reviewers, who helped to shape the book (particular thanks are due Ira Jacobs and Brad Kummer whose assistance in incorporating manufacturing and system details was invaluable), to the students who supplied me with useful feedback, and to Chris Scussel, who aided with the typesetting software. Thanks are also due to Ilona Jones who created the cover (the nature of which will be clear to any reader), originally for the January/February, 1987 issue of the AT&T Technical Journal, and to Andy Meyers, Editor of that publication, who permitted its use.

Finally, I wish to express my appreciation to my other colleagues who good-naturedly endured the preoccupation of their coworker while this text was in preparation.

B. E. B.

CONTENTS

ILLUSTRATIONS LIST

CHAPTER 1

INTRODUCTION TO FIBER OPTICS

1.1 INTRODUCTION

The subject of fiber optics systems did not exist twenty years ago. In the intervening period, what can now be called the *field* of Fiber Optics has grown from a speculation on the possible potential performance of fiber fashioned of the "right" material [1], to many millions of dollars worth of equipment being employed in the field, replacing metallic conductor equivalents or serving in activities transcending the capabilities of prior transmission media. AT&T, for example, has manufactured and shipped more than 10^9 meters of cabled fiber.

Such a rapid growth can be explained in two ways: the new technology has proven more economical than conventional methods in some key applications, and in others its peculiar properties are so desirable that cost is not a significant consideration.

As a one-to-one replacement for most metallic facilities, fiber, at least at present, proves to be more expensive on a per unit length basis (there is, however, considerable potential for cost reduction in manufacture; e.g. continuous processes). On a per hertz of bandwidth per unit length basis, however, fiber has the potential for extremely low cost *in those applications where the available bandwidth can be utilized.* Alternatively, in some applications the extra bandwidth may be traded off against some other desirable property (e.g., immunity to nonlinearities or improved signal-to-noise ratio) to derive equivalent value from it.

By way of example, the bandwidth-distance product of a reasonable-cost graded-index fiber can be of the order of one gigaHz-kilometer per wavelength transmitted, and it has been speculated that it may soon be possible to transmit many distinct wavelengths over such a fiber with adequate crosstalk performance [2] (transmission of 10 wavelengths over a single fiber has already been demonstrated [3]). The bandwidth of single-mode fiber can be two or more orders of magnitude greater. Further, *coherent* transmission techniques offer the potential for yet greater bandwidth utilization.

It will be recognized, of course, that the economies that such enormous bandwidth offers cannot be realized for many applications because their bandwidth needs are much more modest, and such uses must await the availability of lower-cost fiber systems.

Quite different are the applications that can benefit from one or more of optical fiber's unique characteristics. Perhaps most important among fiber's unusual attributes is its essentially absolute immunity to electromagnetic noise ingress. Unlike conventional metallic facilities, which are potentially heir to all coupled electromagnetic interference mechanisms (though careful balance and twisting can eliminate much noise in wire pairs), glass fibers experience no induced, transmittable interfering power coupling from conventional sources, and offer similar immunity to light sources when appropriate sheathing is employed.

Extreme examples of applications of this property are in communications links associated with high-voltage transmission facilities which create large transients [4], and telemetry channels associated with instrumentation for underground nuclear detonations which produce the well-known Electro-Magnetic Pulse (EMP) phenomena [5]. More commonplace applications such as long-distance transmission and intracomputer interconnection are, however, fortunate beneficiaries of this property as well.

Optical fiber is so small in diameter, that, properly sheathed, it is extremely light weight. It is thus a candidate for applications where weight is at a premium, such as a replacement for the many pounds of copper used in present-day aircraft wiring for monitoring, communications and control. The McDonnell Douglas AV-8B or Harrier II V/STOL (vertical/short takeoff and landing) close-support light-attack airplane was, for example, the first aircraft in its class to employ fiber optics [6, 7], and Boeing Co. has announced plans to incorporate a fiber optics flight-control system in its high-technology 7J7 commercial aircraft primarily for weight savings [8].

Present applications of fiber optics systems include telephone trunks [9, 10], data links [11], satellite antenna video entrance links [12], avionics [13], CATV (Community Antenna Television) transmission [14], underwater cable [15, 16], Local Area Networks [17, 18], military applications [19], etc. In addition, optical fiber is employed in fashioning sensors [20], switches [21], modulators [22], etc.

There is some evidence that the future may see the end of the dominance of metallic transmission facilities in major applications. Work is afoot, for example, that takes aim with fiber optics at the present-day copper wire twisted-pair telephone loop plant [23], a very high-volume user of copper. New trunking facilities between telephone central offices are already predominately utilizing optical fiber [24]. Metallic submarine cable transmission facilities are unlikely to ever be deployed again over major (e.g., transatlantic and transpacific) routes.

1.2 REVIEW OF CONVENTIONAL TRANSMISSION MEDIA

Conventional means for transmission are well-known and have been exhaustively studied. Only a few of the attributes of these media will be reviewed for comparative purposes. One point to observe is that bandwidth has historically been viewed as a limited and expensive commodity to be conserved at the expense of end-electronics; indeed, new techniques for conserving bandwidth are still under intensive study [25]. In many applications where fiber is (or soon will be) a reasonable candidate, however, bandwidth may be considered to be essentially unlimited. (Thinking in such terms is a novelty to engineers experienced in conventional transmission system design.)

Similarly, with conventional transmission facilities, attention must be paid to cross-talk and noise ingress and egress over the span. The ability provided by fiber to ignore such concerns is virtually revolutionary in many applications.

METALLIC

Historically, metallic transmission facilities began use for telegraphy in about 1820 with single-wire, ground return systems [26]. Such systems were adequate for many years for telegraphic transmission and were quite economical, but their use for voice transmission after the invention of the telephone in 1876 quickly identified severe limitations due to induced noise from atmospheric, and eventually, power-line sources [27].

Practical, modern metallic transmission systems of significant length are limited to facilities utilizing wire-pair, coaxial cable, and waveguides.

WIRE-PAIR

Excluding open-wire line, which is rare except in older, rural areas, wire-pair is typically employed in a twisted configuration to reduce induced differential magnetic interference via compensating adjacent twist coupling, and to preserve a balanced presentation to induced longitudinal interference so that it can be eliminated via passive structures such as center-tapped transformers or active, differential amplifiers.

The bulk of existing wire-pair is employed to convey voice-frequency signals (loosely defined as a 4,000 Hz band, but actually extending from about 200 to 3,300 Hz [28]) for telephonic purposes, but it is employed as well in carrier systems, primarily utilizing digital formats. At frequencies of 1.5 Mbps, for example, repeaters are required at about 1.6 km (1 mile) spacings; for higher frequencies, repeater spacings are correspondingly smaller. Repeaters and their environments (e.g., manholes) are relatively expensive, and are used as seldom as possible.

COAXIAL CABLE

Coaxial cable is widely used in applications where noise ingress and egress are critical. At low frequencies, it is actually inferior to shielded twisted pair, but at high frequencies, its shielding characteristics are superior. It is intrinsically an unbalanced structure in the sense that the metallic projection areas of the conductors differ. Its loss for a typical size is about 2.5 dB per km at 1 MHz, and increases as the square root of frequency. Low-loss structures are physically large.

For typical applications such as CATV, good-quality cable is readily able to provide 400 MHz of bandwidth with about 0.5 km repeater spacings [29]. In addition, it is used for very-long-haul applications (hundreds to thousands of miles) [30].

Intercontinental submarine transmission is an example of an extreme application of coaxial cable, stretching to the utmost the capabilities of this structure (as well as that of repeaters under the constraints of virtual inaccessibility). Cables comprised of several coaxial units have been employed, which have proven capable of immersion under seven miles of seawater, and provide adequate transmission with four-mile repeater spacings.

Submarine cables typically utilize the conductors to provide power to the repeaters as well as to convey signals, sometimes requiring the application of remarkably high voltages at the end points [31].

WAVEGUIDE

Waveguides represent the ultimate in metallic facility performance. Indeed, for circular waveguide operated in the TE_{01} (Transverse electric field [no longitudinal component]) mode, the loss is inversely proportional to the three-halves

power of the frequency, promising very low attenuation for sufficiently high frequencies [32]. Unfortunately, there are practical problems preventing arbitrarily low loss attainment.

One system planned but not adopted for telephonic applications was to use
the band from 40 to 110 GHz with losses of about 1 dB per kilometer and
repeater spacing of 40 km [33]. Economics and the advent of fiber facilities
rendered such systems obsolete.

UNGUIDED TRANSMISSION

Unguided transmission techniques can be partitioned into the categories of terrestrial radio, satellite, and unguided light.

TERRESTRIAL RADIO

Terrestrial radio transmission for telephonic communications purposes is primarily over line-of-sight hops between antennas spaced up to 60 km apart. Frequencies employed range up to 18 GHz (for short-haul), and are limited to the
common-carrier assigned bands. (Private systems may have other frequencies
allocated to them.)

Problems in such systems range from fading effects and rain absorption to
privacy problems (there has been concern for the possibility of espionage
because the signals are not routinely encrypted [34]). The majority of long-haul
telephone traffic in North America is presently handled by radio facilities.

Long-distance radio transmission is also employed, but suffers from reliability problems due to atmospheric disturbances.

SATELLITE

Satellite systems offer single-hop transmission over substantial portions of the
earth's circumference. Geostationary orbits are the standard, and though
authorized spacings about the earth's equatorial plane are becoming smaller,
the 26,500 mile radius circle is becoming "crowded" (in the sense that antenna
resolving power limits reuse of frequencies).

Problems in satellite transmission thus range from physical crowding at the
equatorial circumference to the flight-time delay corresponding to a net
transmission distance of some 45,000 miles (almost twice the circumference of
the Earth); such delays (especially if associated with echoes) can be disconcert-
ing to the telephone user, but are marginally tolerable. Steps are typically
taken to restrict satellite transmission to at most one leg of a transmission path

for voice to prevent compounded delays. Such facilities are, however, quite satisfactory for other than real-time, two-way voice transmission, e.g., one-way television signals and data transmission.

The biannual sun-transit disabling phenomenon is also a unique disadvantage of satellite transmission (emissions from the sun crossing the plane of the equator at the vernal and autumnal equinoxes briefly overwhelm the satellite emanations at the receiving antenna).

UNGUIDED LIGHT TRANSMISSION

When atmospheric occlusion phenomena cannot occur, or continuous transmission path integrity is not required, unguided light systems may be considered. Line-of-sight transmission is normally required, which can limit general applicability.

Such systems have been applied for example [35] in building-to-building university campus data links using low-power light sources. Communication systems capable of such applications over distances of 3 km or more are commercially available [36]. Earth to the moon and back communication utilizing a powerful light source and a passive reflector on the moon has also been accomplished as has aircraft-to-aircraft transmission of up to 100 miles at altitudes of 4,000 to 37,000 ft [37].

Except for such specialized applications, there does not appear to be much future for unguided light transmission.

1.3 INTRODUCTION TO FIBER OPTICS TRANSMISSION

Fiber Optics transmission is uniquely different from metallic facilities transmission. The fundamental entity that conveys power in the light domain is the photon as opposed to the electron in the electrical domain, and unlike the case in wire transmission where it can be shown that the power is conveyed external to the conductor [38], all of the power in a fiber system is conveyed internal to the fiber.

The electromagnetic interference that plagues wire transmission has no effect upon fiber. Changing magnetic flux coupling the fiber induces no electromagnetic force (and if it did, no current would flow through the insulating fiber, nor would the effects of such a current be detected by the device at the receiver). Capacitive coupling of an electrical signal has no effect, nor does conductive coupling.

Thus, the only potential interference source for an optical fiber is light that may couple into it. Such interfering light can, however, be excluded with ease by a sheath that is opaque at that wavelength and/or a jacket that is highly lossy. Figure 1-1 illustrates the relative noise ingress and egress properties of wire-line facilities compared to fiber facilities.

In a metallic wire-line span, crosstalk ingress and egress, random noise due to atmospheric disturbances, power line harmonics, arcing switch contacts, etc. are potential problems over the entire length of the facility. Near-end crosstalk in metallic facilities is strong-signal egress from the transmitting end of one channel to a physically-near receiving-end of another channel. Because the transmitting-end signal is large in amplitude, the power coupled from it to the other channel may compete effectively against a relatively low-level received signal. Noise internal to the receiver is seldom a significant problem in comparison to span-coupled noise.

In a fiber span, as indicated in the figure (Figure 1-1), the span coupling of interfering noise is essentially nonexistent. The electrical signal level at the transmitter is sufficiently robust to have little problem with noise. The receiver, then, is the major focus of concern because the received signal is typically quite feeble. Thermal noise (primarily from a readily identified resistor in the preamplifier) and shot and dark-current noise in the detector must be dealt with carefully in the design. The receiver package must also be carefully shielded from noise ingress.

Thus, it will be noted that the concerns in the design of a fiber-optics facility are quite different from those of a traditional wire-line facility. Noise and crosstalk are a common enemy, but they attack the two system types in different places, and the defenses mustered against them are correspondingly different.

FIBER-OPTICS HISTORY

The capability of a transparent cylinder to guide light has been known since technical antiquity; the child's device of demonstrating a curved stream of water issuing from a container housing a light source [39] displays "bending" of light (see Figure 1-2) via the identical phenomenon utilized in modern fiber: total internal reflection.

This phenomenon was first demonstrated by John Tyndall in 1870. Water has a refractive index of about 1.33 at room temperature, while air has a value essentially equal to that of free space: unity. As will be discussed in Chapter 2, this difference provides the capability of guiding a portion of the light entering

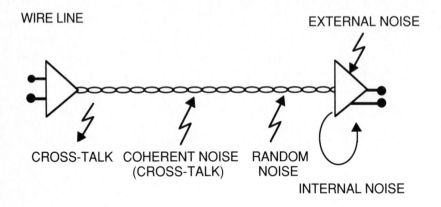

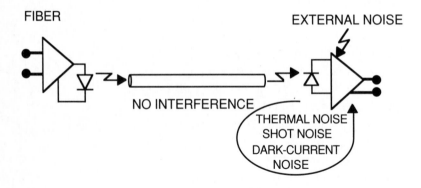

FIG. 1-1. Noise ingress properties: wireline and fiber.

the water stream, even around bends.

Of comparable antiquity is the demonstrated ability to modulate visible light with information-carrying signals of nontrivial frequencies (e.g., voice). Alexander Graham Bell demonstrated the transmission of voice via deflected sunlight for a distance of 200 meters using the *Photophone* in 1880 [40]. Such apparatus was, however, clearly not amenable to adoption as a practical transmission system.

While light sources and receivers are important to the task of light communication, far more critical to the successful fabrication of a practical light-guide system was the parameter of attenuation in the transmission medium

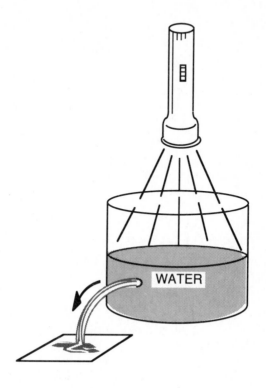

FIG. 1-2. Light guidance by a water stream.

(dispersion, another important parameter, was not as critically important as attenuation, and loomed in significance only after attenuation improvement made fiber light-guide systems of nontrivial lengths possible).

Figure 1-3 shows the history of attenuation improvements in glass, beginning with the Egyptian types which were relatively poor, through the Venetian and Bohemian improvements, to modern window glass, which, though perceptually perfectly transparent to the eye in typical thicknesses, would be very poor as a transmission medium. It is instructive to note the four orders of magnitude difference (in dB) between the transparency of window glass, perceptually lossless, and that required and attained in modern optical fibers.

The step-function in 1970 is due to the advances made in multi-component and high-silica glasses for fiber, and the slower but steady improvements

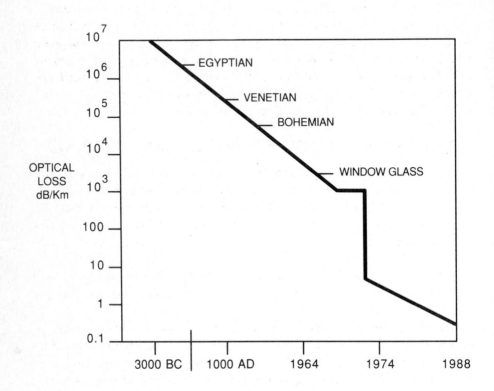

FIG. 1-3. History of glass attenuation.

afterward are principally due to a movement to longer wavelengths, though some refinements and lower doping levels in some fibers have contributed.

A number of materials have received attention as potential guiding media, including liquids, gases, plastics, and, of course, glasses. Glasses embodied the most desirable physical properties for a practical system, but were initially virtually opaque in comparison to the transparency needed for a practical transmission medium (the best fiber available [reexamine the figure] had attenuation of many hundreds of dB per kilometer [41]).

The breakthrough began with the speculation by Kao and Hockham in 1966 [42] that fiber might be fabricated with a loss as low as 20 dB per kilometer, and the observation that fiber with such characteristics could have practical application in telecommunications.

Laboratories in many parts of the world began experiments aimed at approaching the speculated performance. Japanese and British investigators were notably successful in producing low-loss fiber using *multi-component glasses*, but Corning Glass Works in the United States was the first to produce good high-silica fiber (in 1970) [43]. 1970 can thus be regarded as the beginning of the optical communications age, which may one day be viewed as rivaling in significance the advent of the transistor age.

COMPARISON OF METALLIC AND GLASS TRANSMISSION MEDIA

How can metallic transmission facilities best be compared and contrasted with fiber optics systems? The relative advantages and disadvantages can be identified, the unique features can be enumerated, and the collective weight of system experience with the media types can be evaluated. Ultimately, however, the prime consideration is relative cost, that most telling of all parameters (also known as the bottom line).

Certain fundamental observations can be made about the cost per unit of bandwidth of the two system types. The following observations are due primarily to J. H. Mullins [44].

At frequencies of interest for transmission, metallic facilities are limited principally by skin effect, which exacts a loss per unit length proportional to the square-root of frequency, f, and inversely proportional to the circumference (and therefore the diameter, d, of the medium). (Actually, there is typically also a term in the loss equation that is linear in frequency, but at the frequencies of interest for comparison, its contribution is minor [45].) Thus, the loss per unit length, L, can be written as

$$L = k_1 \frac{\sqrt{f}}{d},$$
(1.1)

where k_1 is a proportionality constant. Conductor costs per unit length are proportional to the cross-sectional area of the medium,

$$\$ = k_2 d^2,$$
(1.2)

so that, substituting for d from equation (1.1),

$$\$ = \frac{k_3 f}{L^2} . \tag{1.3}$$

The cost per unit length per unit of bandwidth (specific cost) of metallic facilities, then, taking f as the bandwidth, is given by:

$$\$_m = \frac{k_3}{L^2} . \tag{1.4}$$

In contrast, fiber loss is not dependent upon bandwidth (except for a weak dependency between two classes of fiber with differing bandwidths and core doping levels and therefore losses) or fiber diameter, therefore the cost per unit length per unit of bandwidth, B, is given by

$$\$_f = \frac{k_4}{B} . \tag{1.5}$$

Thus, metallic facilities and fiber have cost dependencies that are not related. For metallic facilities, specific cost is a strong function of the loss; while with fiber, bandwidth (interpreted as *utilized* frequency spectrum) is the sole dependency. Allowable loss in metallic media is limited by the permissible signal to noise ratio, so that costs cannot be driven arbitrarily low, while the bandwidth of a fiber can be very high indeed (though this only has significance in those applications that can usefully employ it).

In practical applications, of course, overall *system* economics must be considered. The costs of end electronics (or the savings thereof over metallic facilities because of increased repeater distances) must be factored in, as must installation costs, the cost (less salvage) of removing old metallic facilities (in replacement applications), etc.

The choice between the two media in a given application for which their system costs are about equal will thus depend upon other criteria. If the span noise is severe, fiber is preferred; if power must also be conveyed to the end user, metal may be preferred; if there is a perceived need in the future for substantial additional bandwidth over the same link, fiber is preferred; etc.

The first major application of fiber facilities was in interoffice telephone *trunks* operated at 45 Mbps, a rate which did not overly strain the bandwidth potential of the fiber, but which constitutes a datum point for utilized bandwidth for the time (1977). Such trunks typically carry significant traffic levels, so their costs may be amortized across a large number of subscribers who (statistically) employ them. Aiding in the economics were the longer repeater spacings afforded by fiber, reducing the number of (and often totally eliminating) expensive manhole-housed, failure-prone *repeaters* [46], and, where replacement took place, the "mining" or salvaging of the old copper. The conversion from copper to fiber for such trunks has been steadily accelerating, and this application is a major user of fiber at this time.

1.4 TRANSMISSION POWER LEVEL DIAGRAMS

It is convenient in consideration of transmission systems to utilize a diagrammatic representation of the power levels (typically expressed in decibels relative to one milliwatt: dBm [see Appendix B]) as a function of the physical position in the system. In later chapters such diagrams will prove useful.

Figure 1-4 illustrates a metallic transmission facility with power levels depicted at each point. Assuming the transmitting amplifier (or *cable-driver*) provides a gain of 10 dB, as does the receiver amplifier, and that the total span loss is 20 dB (e.g., a 10 mile span with 2 dB loss per mile), the diagram depicts the power level at any point. The assumed input signal level is −10 dBm (which corresponds to 0.1 mW).

For wire-line, there is no lossy transduction to and from a different form of signal power, and no excess loss (assuming proper impedance matching) associated with coupling into or out of the transmission line; as will be seen, such is not the case in fiber-optics systems.

1.5 RADIATION EFFECTS

The statement that fiber or fiber systems are immune to electromagnetic interference should be qualified. Certain types of high-intensity interference can have deleterious effects upon the performance of a fiber system. The fiber itself and both the light sources and detectors (as well as the ancillary electronics) can fall prey to high radiation flux levels [47].

The effects can range from *single-event upset*, the losing of a single bit in a digital stream, to longer term and even permanent degradation.

Radiation sources include natural decay processes, artificially generated X-rays, the Cosmos (Cosmic radiation), nuclear power plants, and nuclear device detonations.

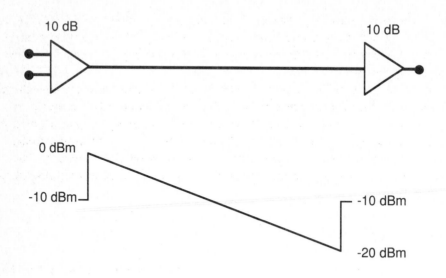

FIG. 1-4. A power-level diagram example.

Particle and ray-producing processes include alpha-decay, beta-decay, positron decay, and gamma ray emission [48].

Alpha decay emits the nucleus of a helium atom or a doubly ionized helium atom. Beta decay causes emission of an electron and an antineutrino. (Both alpha and beta decay typically also involve emission of a gamma ray.)

Gamma rays occur when a nucleus decays from an excited state to its ground state. The energy associated with a gamma ray depends upon the excitation energy of the nucleus. The energies are typically in the range of 0.1 to 10 MeV.

Positron decay emits an electron-like particle with the proper mass but a positive charge: the positron. A neutrino is also emitted.

The term *fluence* is used to describe radiation pulses. It is the time integral of flux and can be in terms of particles per square cm (e.g., photons or neutrons) or energy per square cm [49]. The unit *rad* corresponds to a radiation dose of 100 ergs/gm of irradiated material.

It should be pointed out that normal background radiation is not a problem for conventional fiber optic systems.

Typical sea-level background radiation is estimated at 0.1 rad/year, while spacecraft and nuclear power plant levels may exceed 1 rad/day [50].

Cerenkov radiation is the result of charged particles propagating in a medium at a velocity greater than the phase velocity of light in that medium. The result rather than the means of irradiation, it may interfere with light-borne signals in fiber.

The effects of radiation upon the constituents of a fiber optics system will be discussed in the corresponding chapters, but in brief, with only rare exceptions, irradiation with significant volumes of any species of energetic particle or ray may have deleterious effects on virtually all constituents of a fiber-optic system. There is therefore considerable interest in identifying relatively immune devices and materials, and in operating at wavelengths that may be less affected than others.

1.6 SUMMARY

The field of fiber optics is relatively young. In the span of about 20 years it has become a pervasive yet still rapidly growing technology. Traditional transmission techniques include use of metallic media (wire-pair, coaxial cable, and waveguide) and unguided transmission (terrestrial radio, satellite, and light). Wire-pair is virtually omnipresent in the telephone loop plant, but is severely limited in bandwidth. Coaxial cable is widely used for long-haul (e.g., transcontinental and transoceanic) communication, and for broadband communication (e.g., CATV distribution), but is relatively expensive. Metallic waveguide, though employed for very short-distance microwave frequency transmission, is too expensive in comparison to alternatives, for long-haul usage. Terrestrial radio is widely employed in long-haul telephonic trunking, but it is limited to line-of-sight at microwave frequencies, and is subject to atmospheric disturbances. Satellites are severely limited in number and frequency allocation, and therefore total channel capacity, and introduce a delay disturbing to voice communication. Unguided transmission of light is line-of-sight, and subject to severe atmospheric disturbances.

Fiber optics systems offer new dimensions in bandwidth, size, weight, repeater spacing, and cost/unit of utilized bandwidth, as well as virtual immunity to noise and cross-talk intrusion over the span. Hard radiation, however, can disturb fiber systems by increasing attenuation temporarily or permanently, and by transient disruption of transmission.

Power-level diagrams are useful for depicting system power as a function of spatial position for fiber optic as well as other transmission systems.

EXERCISES

1. Draw a Power-Level Diagram for an end-to-end telephone connection between two telephones each at the end of maximum (1300 ohm loop resistance) loop lengths. (With standard 26 Gauge cable, this corresponds to 3 loop miles and 2.67 dB per mile.)

2. In a suitable handbook, find the attenuation of the largest reasonable-size twisted-pair cable, and determine how much gain would be necessary for coast-to-coast transmission of an analog voice-frequency signal. Compute the weight of the copper involved and the corresponding cost at current copper prices. Compute the weight of a 100 μm diameter fiber of the same length.

3. For a 1 gigahertz-kilometer bandwidth-distance product fiber, how many repeaters would be necessary for a coast-to-coast, 40 MHz channel if attenuation were not a factor? For a 0.25 dB/km fiber, how many repeaters would be necessary if bandwidth were not a factor?

4. How many photons per unit time must be intercepted by a Television antenna in order to deliver one millivolt at 300 ohms (a standard level) to a TV set rf inlet at a typical channel frequency, e.g., 60 MegaHz?

5. The Antares Carbon-Dioxide laser at Los Alamos Laboratory has been reported to have generated 12 trillion watts of power for one nanosecond [51]. How much energy does this represent? How many photons were needed to convey this energy?

6. In 1983 [52], it was announced that an airborne laser operating at 10.6 μm was successful in destroying several sidewinder missiles launched toward it. Make estimates about the size (viewed from the front) and mass of the sidewinder, and calculate how much energy would have to be focused upon it for destruction. How many photons would be involved? How much power would be involved if the irradiation period were one millisecond? How large would the E and H fields be? How much larger could the power level be before spontaneous breakdown of the air would take place and rob the beam of its power?

7. Compare the performance (e.g., bandwidth and repeater spacing) of circular waveguide and fiber. Estimate the relative cost per unit of bandwidth for the two technologies.

8. Compute the altitude necessary for a satellite to be in geostationary orbit around the earth. What must be its linear velocity at this altitude?

9. Identify typical values for k_3 and k_4 as used in Equations 1.4 and 1.5. How much would k_4 change if coherent transmission techniques utilizing all of the potential bandwidth of the light carrier were employed?

REFERENCES

[1] K. C. Kao and G. A. Hockham, "Dielectric Fibre Surface Waveguides for Optical Frequencies," *Proceedings of the IEEE*, vol. 113, July, 1966, pp. 1151-1158.

[2] "Wavelength Multiplex, Demultiplex Devices from Fiber-Optic Research," *Electronics Design News,* February, 1981, pp. 73-74.

[3] "N. A. Olsson, et al., "Transmission with 1.37 Terabit-km/s Capacity Using 10 Wavelength-Division Multiplexed Lasers at 1.5 Microns," *Conference on Optical Fiber Communications*, February, 1985.

[4] E.g., W. M. Hall, et al., "Integration of SCADA and Line Protection for 138 kV Transmission Line Using a Fiber Optic Network," *IEEE/PES Transmission and Distribution Conference and Exposition*, Anaheim, California, September 16, 1986.

[5] W. M. Caton, "Fiber Optics Systems in Adverse Environments, II," *Proceedings of the International Society for Optical Engineering,* August, 1981.

[6] C. C. Sanger and D. L. Williams, *Fibre Optics 1983,* April 19-21, 1983, p. 141.

[7] "Fiber Optics Proved in Harrier," *AVIONICS*, March, 1985, p. 14.

[8] R. W. Lay, "Boeing Using Fiber Optics in New Jets," *Electronic Engineering Times*, February 10, 1986, p.6.

[9] I. Jacobs and J. R. Stauffer, "FT3 - A Metropolitan Trunk Lightwave System," *Proceedings of the IEEE,* vol. 68, No. 10, 1980, pp. 1286-90.

[10] D. Bly, "Fiber is Cost-Effective Now," *Telephone Engineer and Management,* vol. 89, no. 1, January, 1985, p. 52.

[11]   P. B. O'Connor et al., "Large Core High NA Fibers for Data Link Applications," *Progress in Optical Communication,* IEE Reprint Series 3, p. 196.

[12]   A. Albanese and H. F. Lenzing, "IF Lightwave Entrance Links for Satellite Earth Stations," *IEEE International Communications,* 1979, p. 1.7.1.

[13]   G. Belcher and D. Marshall, "Use of Fiber Optics in Digital Automatic Flight Control Systems," *IEEE Transactions on Aerospace Engineering Science,* September, 1975, pp. 841-850.

[14]   M. Kawahata, "Fiber Optics Application to Full Two-Way CATV System - HI-OVIS," *National Telecommunications Conference Record,* 1977, p. 14:4.

[15]   C. D. Anderson, R. F. Gleason, P. T. Hutchison, and P. K. Runge, "An Undersea Communication System Using Fiber-Guide Cables," *Proceedings of the IEEE*, vol. 68, no. 10, 1980, pp. 1299-1303.

[16]   P. Worthington, "Cable Design for Optical Submarine Systems," *IEEE Journal on Selected Areas in Communications, Undersea Lightwave Communications*, November 6, 1984, pp. 873-878.

[17]   M. Hirai, W. Kawase, H. Kobayashi and Y. Katsuyama, "Optical Fiber Cables for Local Area Network," *IEEE International Conference on Communications (ICC '83)*, June 1983, pp. 707-712.

[18]   D. J. Cunningham and S. S. Lymphany, "Ethernet: Fiber Optic Design Issues and Applications Experience," *FOC/LAN 84*, September, 1984, pp. 101-108.

[19]   M. K. Barnowski, "Fiber Systems in the Military Environment," *Proceedings of the IEEE*, vol. 68, no. 10, 1980, pp. 1315-1320.

[20]   H. F. Taylor, T. G. Giallorenzi, and G. H. Siegel, Jr., "Fiber Optic Sensors," *First European Conference on Integrated Optics,* September, 1981, pp. 99-100.

[21]   Y. Fujii, et al., "Low-Loss 4X4 Optical Matrix Switch for Fibre-Optic Communication," *Progress in Optical Communication,* IEEE Reprint Series 3, pp. 184-185.

[22]   R. C. Alferness, C. H. Joyner, and L. L. Buhl, "High Speed Traveling-Wave Directional Coupler Modulator for Lambda = 1.32 Microns," *1983 Optical Fiber Communication Meeting, Optical Society of America*, p. 20.

[23]   K. Y. Chang, "Fiberguide Systems in the Subscriber Loop," *Proceedings of the IEEE*, vol. 68, no. 10, 1980, pp. 1291-1299.

[24]   E. E. Basch, R. A. Beaudette, and H. A. Larnes, "Optical Transmission for Interoffice Trunks," *Transactions of the IEEE Communications Society*, vol. 26, no. 7, 1978, pp. 1007-1014.

[25]   E.g., B. S. Atal and R. Q. Hofacker, Jr., "The Telephone Voice of the Future," *AT&T Bell Laboratories Record*, July, 1985, pp. 4-10.

[26]   E. Hawks, *Pioneers of Wireless*, Methner & Co. Ltd., pp 60-86.

[27]   B. E. Briley, *Introduction to Telephone Switching*, Addison-Wesley, 1983, p. 2.

[28]   Bell Laboratories Staff, *Transmission Systems for Communications*, Bell Telephone Laboratories, Inc., 1971, p. 39.

[29]   J. R. Fox, D. I. Fordham, R. Wood, and D. J. Ahein, "Initial Experience with Milton Keynes Optical Fiber Cable TV Trial," *IEEE Communications*, September, 1982, p. 2155.

[30]   *Transmission Systems for Communications*, p. 305.

[31]   *Reference Data for Engineers*, H. W. Sams, 1975, pp. 35-36.

[32]   A. J. Baden Fuller, *An Introduction to Microwave Theory and Techniques*, Pergamon Press, 1979, pp. 135-39.

[33]   Bell Laboratories Staff, *Engineering and Operations in the Bell System*, Bell Telephone Laboratories, Inc., 1977, pp. 340-43.

[34]   "Soviets Accused of Listening-In on Phone Calls," *Chicago Tribune*, Friday, January 13, 1984, p. 5.

[35]   Y. Ueno and R. Nagura, "An Optical Communication System Using Envelope Modulation, *IEEE Transactions on Communications*, vol. COM20, no. 4, August, 1972.

[36] E.g., the GOALS system offered by General Optronics Corporation of Edison, NJ.

[37] "Warplanes 'Talk' via Laser Beam," *Machine Design*, June 12, 1986, p. 16.

[38] E.g., D. K. Cheng, *Field and Wave Electromagnetics*, Addison-Wesley, 1983, p. 326ff.

[39] J. Hecht, "Victorian Experiments and Optical Communications," *IEEE Spectrum*, vol. 22, no. 2, February, 1985, p. 69.

[40] A. G. Bell, "Selenium and the Photophone," *The Electrician*, vol. 5, 1880, p. 214f.

[41] T. Li, "Advances in Optical Fiber Communications: An Historical Perspective," *IEEE Journal on Selected Areas of Communication*, vol. SAC-1, no. 3, April, 1983, pp. 356-72.

[42] Kao.

[43] F. P. Kapron, D. B. Keck, and R. D. Maurer, "Radiation Losses in Glass Optical Waveguides," *Applied Physical Letters*, vol. 17, 1970, pp. 423-425.

[44] H. Kressel, *Topics in Applied Physics*, vol. 39, Springer-Verlag, 1980, pp. 260-261.

[45] A. S. Taylor, "Characterization of Cable TV Networks as the Transmission Media for Data," *IEEE Journal on Selected Areas in Communication*, vol. SAC-3, no. 2, March, 1985, p. 259.

[46] D. Michalopoulos, "Fiber Telecommunications System Reduces Need for Signal Regeneration," *Computer*, March, 1982, p. 96.

[47] J. E. Gover and J. R. Srour, *Basic Radiation Effects in Nuclear Power Electronics Technology*, Sandia Report: SAND85-0776, May, 1985.

[48] Ibid, p. 14ff.

[49] Ibid, p. 31.

[50] E. I. Friebele, "Radiation Effects in Optical Communications Systems: Fibers Exposed at Low-Dose Rates and Selfoc Lenses," *Technical Digest, Conference on Optical Fiber Communications*, New Orleans, LA,

January, 1984, p. 102.

[51] *Machine Design*, vol. 55, no. 25, November 10, 1983, p. 10.

[52] "Laser Blasts Missiles," *Machine Design*, vol. 56, no. 23, October 11, 1984.

CHAPTER 2

OPTICAL FIBER

2.1 PRELIMINARIES

Modern-day optical fiber is primarily composed of high-silica glass doped with certain impurities that alter the refractive index, n, which is the ratio of the speed of light in a vacuum to the speed of light in the substance of interest. This number is, of course, always greater than or equal to one.

The index of refraction is, in general, complex, separable into real and imaginary components, the latter of which is sometimes called the *extinction coefficient* [1] because it relates to the signal strength decay rate or attenuation. The real and imaginary components of a response function (corresponding here to dispersion and attenuation, respectively) are interdependent in a manner directly analogous to the observations of H. Bode for electrical systems [2] (as pointed out by Gowar [3]). Over wavelengths of interest to transmission, however, the attenuation must be low (away from atomic and electronic resonances), so that the imaginary part of the index of refraction may be considered negligible compared to the real part.

Figure 2-1 displays a short section of optical fiber of a variety known as *step-index*. In such fiber, the refractive index undergoes an abrupt change from a value in the center portion or *core* to a slightly (of the order of one percent) lower value in the outer portion of the cylinder. (The lower index of refraction outer portion of the fiber is referred to as the *cladding*, not to be confused with the protective *jacket* that is usually placed upon the fiber over the cladding just after it is formed during manufacture to prevent mechanical damage and water vapor absorption.)

The property of refraction at a surface is also displayed in the figure. It will be recalled that refraction is governed by Snell's Law:

$$n_1 \sin \phi_1 = n_2 \sin \phi_2 \qquad \text{SNELL'S LAW}, \qquad (2.1)$$

where the ϕs are angles to the normal on either side of the interface between two homogeneous, isotropic media (see Figure 2-2), and the indices of refraction are those of the material on either side.

When ϕ_2 reaches 90 degrees, none of the light impinging upon the interface is transmitted through it, and *total internal reflection* takes place. The value of ϕ_1 at this point is referred to as the *critical angle* (given by $\phi_{crit.} = \sin^{-1} n_2/n_1$), and light impinging at or above this angle will, if there is a corresponding parallel interface to material with the same characteristics, be *guided*. (As will be seen later, this is a simplification of a more complex phenomenon; the redirection of light does not take place at an interface of vanishingly small dimensions, and, in general, a portion of the light penetrates into the cladding.)

2.2 COMPOSITION

Light may be guided by a variety of substances, ranging from cylinders of liquids [4] or gases [5], to regular solids of materials of appropriate transparency such as plastics and glasses. High silica glasses at present possess the most desirable characteristics for commercial transmission applications. (A later chapter will discuss the category of wave-guides comprising, e.g., titanium diffusion into lithium niobate ($LiNbO_3$), which takes advantage of the increase in refractive index caused by the diffused metal.)

Most commercial fibers are manufactured of silica (SiO_2) doped with appropriate oxides (typically germanium, fluorine or phosphorus) to alter the index of refraction. Alternative materials and dopants with improved properties (e.g., with respect to immunity to nuclear radiation effects) and/or lower cost are also under study [6].

2.3 MANUFACTURE

There exist several processes for the manufacture of optical fiber: the Outside Vapor Deposition (OVD) technique, the Modified Chemical Vapor Deposition (MCVD) technique, the Vapor Axial Deposition (VAD) technique, and the Double Crucible (DC) technique. A number of variations upon these basic processes also exist [7].

Fiber may also be fabricated of plastics, typically polymethyl-methacrylate, which can utilize the surrounding air or jacketing as a cladding. Plastic

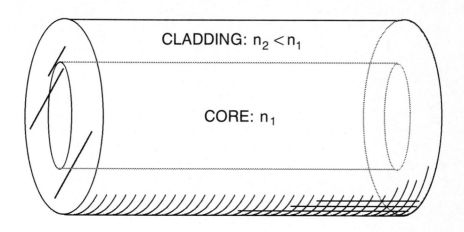

FIG. 2-1. Step-index fiber.

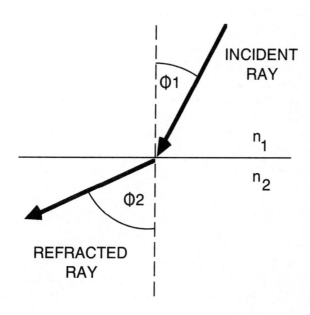

FIG. 2-2. Snell's law.

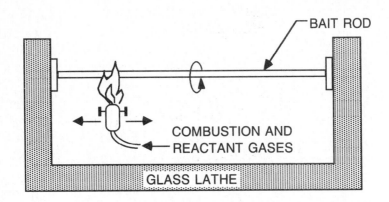

FIG. 2-3. The OVD process.

cladding can also be placed upon a silica core.

The all-plastic fiber is transparent principally at visible wavelengths, requiring corresponding sources and detectors. The attenuation in plastic-clad silica fiber is significantly higher and bandwidth lower than that of all-silica fiber.

All-plastic fiber is an economical medium for use in short-run applications (its attenuation is of the order of hundreds of dB per km) at low modulation frequencies (the fiber bandwidth is typically low, and visible light sources usable at very high frequencies are not readily available). Plastic coated silica is a cheaper variety of fiber than all-silica types, since the cladding coating can become identically the jacket.

Both types of plastic fiber have problems under extreme temperature conditions, the plastic softening at elevated temperatures (e.g., 150°C, not unknown under storage conditions), and embrittling at very low temperatures (e.g., −55°C, a known avionics extreme). At reasonable ambients, however, reported longevity results with plastics have been good enough for some applications.

THE OVD TECHNIQUE

The OVD process was pioneered by Corning [8], and was used to make some of the earliest high-quality fiber.

The OVD process begins with a solid cylindrical "bait-rod" [9] of graphite or ceramic mounted in a lathe and rotated while a *soot* of doped glass is

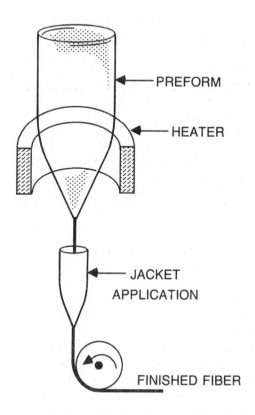

FIG. 2-4. Fiber drawing.

deposited upon the outside via a flame oxidation process at a temperature of 1300–1600°C (see Figure 2-3). The first layers deposited will become the core, and the last layers, the cladding. (A varying index of refraction is also possible with this process.) The bait-rod is removed after processing, and the porous soot boule remaining is subjected to one or more processes to remove OH radical contamination while being sintered into continuous glass to produce a *preform*.

The preform (for this and other processes to be discussed) is then placed in a vertical position and heated to about 2000°C at the lower end, allowing the flow of molten glass (see Figure 2-4). By controlling the temperature and the force pulling upon the glass, a uniform fiber with a diameter as small as 125

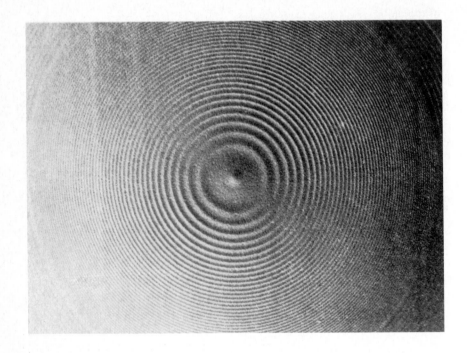

FIG. 2-5 Preform Profile.

μm can be routinely drawn. A preform one meter in length may yield some 20 to 40 kilometers of fiber.

The fiber (for this and other manufacturing processes) is typically protected from mechanical damage and water vapor intrusion by immediate coating with a durable plastic *jacket*. (The choice of jacketing material can be critical: in 1984, evidence of in-place fiber degradation was noted and the culprit was determined to be a jacket material that evolved elemental hydrogen [10]. Care in the choice of jacketing material can prevent this problem, to which all of the fiber types to be discussed are potentially sensitive.)

One of the most remarkable consequences of the drawing process is that the refractive index profile of the fiber is (ideally) a faithful reproduction of that of the preform (see Figures 2-5 and 2-6). It is thus possible to infer some of the critical properties of the fiber to be drawn from those of the preform, allowing elimination of flawed material before the expense of the drawing process. Extensive tests are in fact typically performed upon preforms before the drawing process.

THE MCVD TECHNIQUE

The MCVD technique developed at Bell Laboratories (now AT&T Bell Laboratories) [11] starts with a hollow cylinder of pure silica which is mounted upon a glass-lathe (see Figure 2-7); the cylinder material serves as much of the cladding for the resulting fibers (some of the cladding material is typically deposited for lowest possible loss). Gasses comprised of the constituents of materials to be deposited are passed through the tube while a oxy-hydrogen flame is repetitively passed over the rotating cylinder, promoting oxidation and the deposition of a soot of the combined reactants upon the *inner* surface of the tube, which builds up in successive layers. Note that this process permits the deposit of layers of varying composition in a carefully controlled atmosphere.

When a sufficient thickness of material has been deposited, typically requiring two to three hours of continuous deposition, the tube is heated and permitted to collapse into the solid cylindrical preform; the collapse is the result of surface tension in the softened glass. Drawing is then performed as described above.

Deposition speeds have been accelerated via techniques that support a high-temperature plasma within the tube, which is fed power by an RF source. Deposition rates have been increased by a factor of as much as four [12] using this technique.

THE VAD TECHNIQUE

The VAD technique, developed by Nippon Telegraph and Telephone (NTT) [13], produces a preform via axial deposition of the desired materials (see Figure 2-8). (This technique has some similarity to the OVD process.) Because the processing may not take place within as well protected an atmosphere as does the MCVD process, the quality of the fiber produced can be more difficult to control. Nevertheless, fibers with remarkable properties have been produced with this process.

The possibility of very economical fiber may be embodied in processes such as VAD because of their (thus far largely unrealized) potential for continuous manufacture. A variation upon the process announced by NTT called the "Twin-Flame High-Speed VAD" technique holds promise of a decimal order of magnitude increase in manufacturing speed [14].

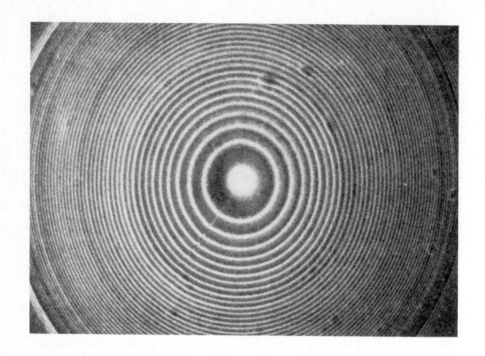

FIG. 2-6. Fiber profile.

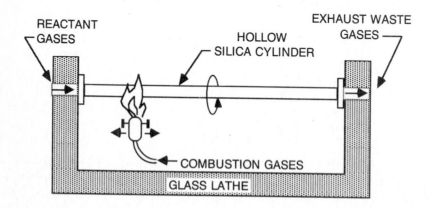

FIG. 2-7. The MCVD process.

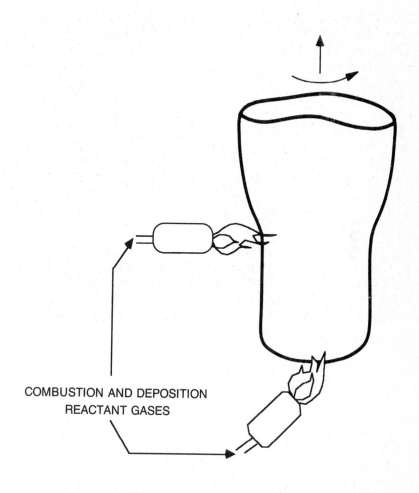

FIG. 2-8. The VAD process.

THE DOUBLE-CRUCIBLE TECHNIQUE

The Double Crucible method provides the core and the cladding materials in separate crucibles, combining them physically in the pulling process (see Figure 2-9) [15]. This method to date has not proven capable of producing high-quality fiber, but it too is a potentially continuous method [16], and refinements may improve the quality of its output. The choice of crucible material is critical to the production of pure glass because the crucible itself is a potential source of contaminants.

The Double Crucible method has not yet progressed beyond the laboratory or pilot-line stage because of practical difficulties.

2.4 DIMENSIONS

Typical fiber will have core diameters in the range of about 8-10 μm for single mode fiber and 50-200 μm for multimode fiber. The outer cladding diameter is typically 125 μm (standards include Core/OD: 8/125 μm - depressed-cladding single-mode; 10/125 μm - matched-cladding single-mode; 50/125 μm, 62.5/125 μm, and 85/125 μm - graded index). The protective overlayer jacketing will increase the physical size of the fiber by several tens of μm more (250 μm is a standard overall diameter).

Fiber cable, which includes strengthening members and sometimes ancillary conductors as well as a tough abrasion and water resistant sheath, may be as small as microcoax for single fibers, and as large as an inch or so in diameter for several hundred fibers [17] (see Figure 2-10, which depicts cable capable of accommodating 144 fibers in a layered structure providing moisture resistance and mechanical strength with an overall outside diameter of about one-half inch [12 mm]). An important additional function of such cables is to limit bending radius to protect the fibers. Special-use cables such as submarine types tend to be relatively massive because of the need for special strengthening.

2.5 CHARACTERISTICS

The characteristics of fiber that are of interest to the designer range from its transmission parameters to its mechanical properties, all of which are unique in comparison to wire transmission.

ATTENUATION

The most significant attribute of modern optical fibers is its extremely high transparency (or low attenuation) in certain wavelength ranges. Early fiber was considered excellent if it displayed 10 dB per kilometer attenuation at usable wavelengths. Today, fiber is readily available displaying attenuations ranging from 5 to 0.2 dB per kilometer, depending upon fiber type and wavelength, and experimental results yielding losses of 0.16 dB per kilometer have been obtained [19]. Fibers (as yet unrealized in practical form) utilizing alternative

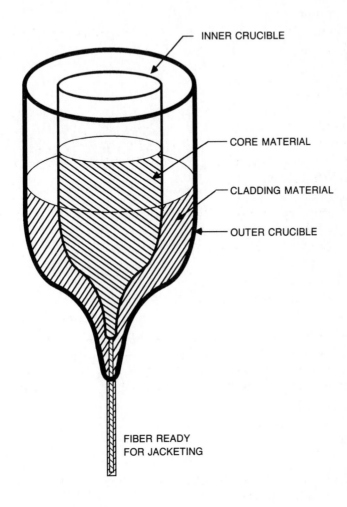

FIG. 2-9. The Double-Crucible process.

materials have been predicted to yield attenuation as low as 0.001 dB per kilometer [20].

Figure 2-11 displays the attenuation characteristics of typical graded-index fiber. It will be noted that the attenuation is a strong function of wavelength, and that there exist resonances comprising harmonics of hydroxyl ion (water) vibrations. Indeed, water proved the major attenuation cause in early fiber, and

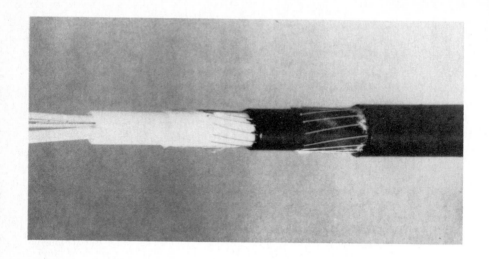

FIG. 2-10. Fiber cable.

only when manufacturing refinements were instituted did it become a near-negligible consideration over much of the spectrum of interest.

Figure 2-12 displays the attenuation characteristics of modern single-mode fiber. Note the improvement attributable primarily to the reduced dopant concentrations in the core.

On an experimental basis, some fiber has been produced with the characteristics of Figure 2-13, essentially eliminating water as an influence in attenuation [21].

The earliest light launching and receiving devices were only useful at short wavelengths of the order of 0.8 μm, but devices are now being made which permit operation essentially anywhere of interest in the fiber "window" bounded by infrared absorption and by a scattering effect (Rayleigh) which will shortly be discussed. Further, it is possible to operate simultaneously at a multiplicity of wavelengths (wavelength division multiplexing) without appreciable interference, greatly multiplying the usable bandwidth of a fiber.

There are two major constituents to the net attenuation of a fiber: absorption and scattering.

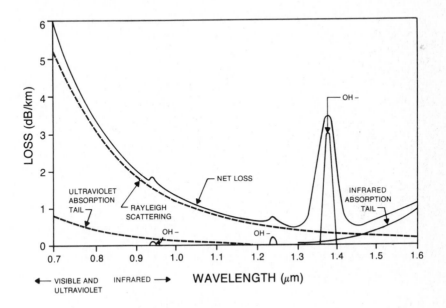

FIG. 2-11 Attenuation characteristics of graded-index fiber. (Reproduced with permission from Jefferies and Klaiber, [18].)

Absorption

Absorption of light power in high-silica glasses is due to intrinsic wavelength-dependent losses in the ultraviolet and infrared as well as the overtones of the fundamental hydroxyl (OH) vibration at 2.72 μm. These OH peaks occur at about 0.72, 0.95, and 1.38 μm. Additional absorption peaks occur at combinations of these frequencies and the vibration frequencies of the SiO_2 tetrahedron at 0.88, 1.13, and 1.24 μm [23]. In current fiber, only the 1.38 μm peak is of significance.

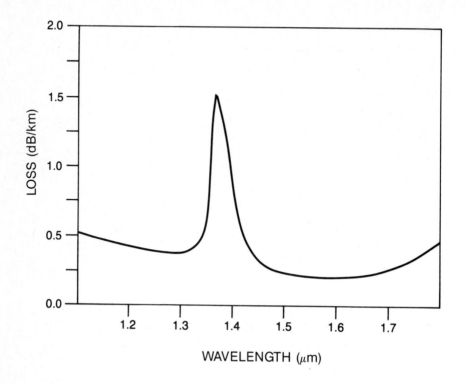

FIG. 2-12. Example attenuation characteristics of single-mode fiber. (Adapted from data supplied by R. B. Kummer, AT&T Bell Labs.)

Scattering

Scattering takes place as a result of compositional fluctuations in the fiber, some subject to reduction by careful control of the manufacturing process, and some intrinsic to the material.

- Rayleigh Scattering

  Rayleigh scattering is due to the irreducible randomness of the molecular structure of the glass. It is inversely proportional to the fourth power of the wavelength:

$$\alpha_R = \frac{k}{\lambda^4} .$$

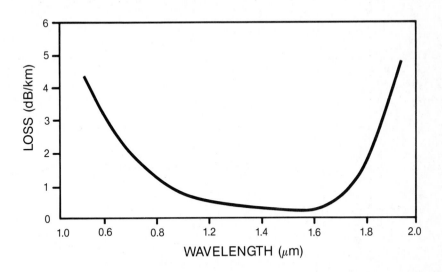

FIG. 2-13. VAD-produced, very-dry fiber. (Reproduced with permission from Moriyama et al., [22].)

The proportionality constant is a function of the density of impurity dopants employed.

It is primarily Rayleigh scattering that prevents effective utilization of wavelengths significantly below 0.8 μm in silica fiber.

• Mie Scattering

Mie scattering is caused by relatively massive imperfections in the glass such as bubbles or microcracks. This scattering can be reduced through careful control of the manufacturing process.

• Waveguide Scattering

Waveguide scattering is the result of variations in core diameter and imperfections in the core-cladding interface which cause departures from the perfectly cylindrical. Transfer of energy to radiation modes is the loss mechanism.

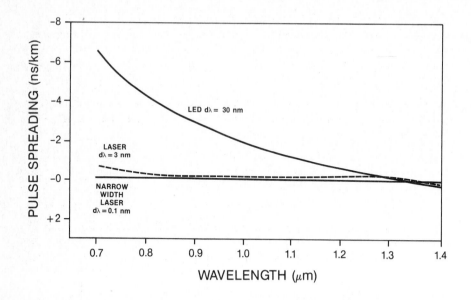

FIG. 2-14. Material dispersion for pure silica. (Reproduced with permission from Jefferies and Klaiber, [24].)

Bending and Microbending

Sharp bends (macrobends) can contribute to loss as well as threaten the integrity of the fiber. Typically the cable design will protect the fiber from excessively tight bends.

It is obvious from ray optics considerations that a bend in a fiber guide will cause some high-order mode rays (when they exist) to impinge on the core-cladding interface at less than the critical angle and become lossy, while low-order modes will be unaffected. Other considerations, however, predict correctly that all modes will lose power at a bend [25]: the wavefront propagating in a fiber will be compelled to move faster at the outer periphery of the bend than at the center of the fiber; it would be necessary to exceed the speed of light in the material to accomplish this feat, so that loss must occur; since all modes are represented in the wavefront, all lose power at a bend.

Ray optics are not useful for analyzing the behavior of light in single-mode fiber, but consideration of the power distribution (mode field) and its effective dimensions (mode-field diameter [to be defined shortly]) is useful for computation and visualization of the effects of bending and microbending in single-mode fiber. The degree of confinement of the mode field is diminished by macro and microbending, which effectively distort the index profile [26] and promote leakage of power. So long as the bending does not exceed reasonable limits (e.g., a loop diameter of about 5 cm), the loss will typically be negligible, and the technique of providing a depressed cladding structure (to be described shortly) can allow tighter bends because the mode field is more effectively confined [27].

Mode field radius (or diameter, as is more often quoted) is a strong function of the wavelength, increasing faster than linearly with λ [28] (one term in the expression is proportional to λ^6). Because bending losses are a function of the mode field diameter in single-mode fiber, they also depend upon λ (this is in contrast to multi-mode fiber, where bending losses are essentially independent of λ). In particular, macrobending loss is an exponentially increasing function of λ, while microbending loss may increase (the typical case) or decrease with λ [29] depending on the axis deformation function.

Microbends are microscopic, short-period, random bends typically impressed upon the fiber during the manufacturing process. Microbending losses can be significant, but attention to good manufacturing practice and the use of compliant coatings can improve upon this phenomenon. Some residue of microbending always remains however, and considerations such as depressed cladding structures for single-mode fiber become important when this loss component must be minimized. Microbending loss sensitivity is strictly dependent upon mode field diameter for single-mode fiber, and the smaller this value (typically $\approx 9 \ \mu m$), the lower the microbending loss can be expected to be.

DISPERSION

Dispersion is the mechanism which limits the bandwidth of the fiber. It is the result of either a wavelength-sensitive effective propagation velocity which causes, for example, a pulse of light composed of a multiplicity of wavelengths to arrive dispersed in time (material dispersion), or a geometrical, flight-path length difference between elements of light, even if at the same wavelength, causing them to arrive at the receiving end at different times if they entered the fiber at different angles (modal dispersion).

• Material Dispersion

Material or "chromatic" dispersion is an intrinsic material property which is a function of wavelength. Figure 2-14 illustrates the characteristics of this property for pure silica. Note, in particular, that at about 1.3 μm (close to a region of minimum attenuation), this source of dispersion goes to zero for silica fiber. Because the lowest attenuation is in the 1.5 - 1.6 μm range, there has been much interest in shifting the dispersion zero to that region.

Material dispersion is more profound when the light source has a broad spectrum such as that characteristic of Light Emitting Diodes (LEDs) (typically 30 to 100 nm between half-power points). Injection Laser Diodes (ILDs), in contrast, have very narrow spectra (typically 3 nm), and their emissions are consequently much less affected by material dispersion. Light launched by very high-quality, single-longitudinal-mode lasers which produce extremely narrow spectra (e.g., 0.1 nm) is virtually immune to this effect.

It is possible to adjust doping profiles to shift the zero-dispersion wavelength to produce *dispersion-shifted* fiber [30], and even to induce two zero-dispersion points [31], widening the low dispersion region (*dispersion-flattened* fiber).

• Waveguide Dispersion

Another wavelength-dependent dispersion mechanism is waveguide dispersion, which is due to the wavelength dependence of modal group velocity [32]. It is possible to shift the zero dispersion wavelength somewhat by trading off material dispersion against waveguide dispersion.

• Modal Dispersion

Modal dispersion is best explained diagrammatically (see Figure 2-15). It is clear from the figure that light entering the fiber face off-normal will experience a flight-time greater than light entering on the normal, which will describe the shortest path through the fiber. Depicted are the extremes, the axial ray and the ray entering at the limit of the *acceptance cone* and taking the most tortuous path. Clearly, this phenomenon is manifest only in multi-mode situations; single-mode fiber is immune to modal dispersion.

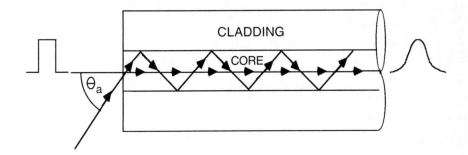

FIG. 2-15. Modal dispersion.

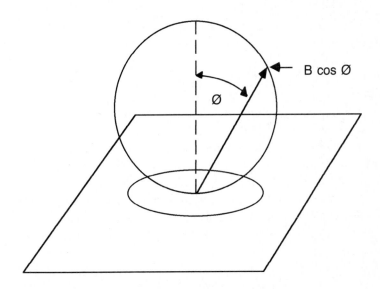

FIG. 2-16. Typical surface-emitting LED radiation pattern.

As was true for material dispersion, this effect might be expected to be more severe for LEDs than for ILDs because the former's radiation patterns are broader (e.g., for surface emitting

LEDs, the pattern is approximately *Lambertian* [with an amplitude as a function of the angle from the normal proportional to the cosine thereof] see Figure 2-16). In practice, however, both source types can excite all modes in a multi-mode fiber, and mode mixing at splices and connectors tends to further reduce the distinction between the effects of source types. It should be emphasized, however, that material rather than modal dispersion usually dominates with LED sources at short wavelengths.

Fortunately, an important technique called index grading has been developed which greatly enhances the bandwidth of multi-mode fiber. This technique introduces a refractive-index gradient in the core which makes the *flight-time* much less dependent upon the distance (optical path length) traveled by a light packet in the fiber (see Figure 2-17).

Though, as in ungraded (step-index) fiber, rays entering the fiber off-normal traverse a lengthier path, more of the path is in material of lower refractive index, so that the light travels more rapidly in it. In an ideal graded-index fiber, the flight-times for all possible paths would become equal, and no modal dispersion would result.

Ideally, a smoothly varying refractive index would be employed (and has been achieved via ion exchange processes [33]), but techniques such as the MCVD process achieve comparable results by depositing discrete layers of progressively higher refractive index. Such a discrete layering is helpful conceptually in visualizing the light bending effect that is produced: many discrete refractions can be visualized as taking place in the layered glass, leading to a redirection of the light. A continuously changing index can be viewed as allowing the number of layers to increase without bound. (In practice, diffusion during deposition and drawing causes discrete-layered fiber to evidence a virtually smooth refractive-index profile.) A parabolic index profile can be shown to be near ideal.

The effect of grading is analogous to that experienced in atmospheric transmission over a wide band of radio frequencies as a result of the ionospheric E and F layers [34].

The improvement in bandwidth from the use of graded index fiber is profound, typically amounting to two decimal orders of magnitude. This fiber type will be discussed in more detail shortly.

Because single-mode fiber is free of modal dispersion, its dispersion characteristics are far superior to step-index fiber, and much better than even

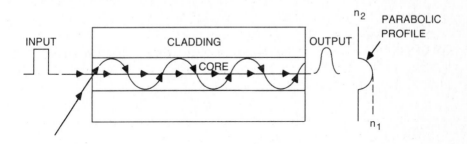

FIG. 2-17. Graded-index fiber.

graded-index fiber (see Fig. 2-18). The corresponding bandwidth behavior of typical single-mode fiber as a function of both wavelength and spectral width of the source is shown in Fig. 2-19.

Single-mode fiber performance is clearly dependent upon the nature of the illuminating source, and knowledge of this fact is important in interpreting fiber manufacturer data. Manufacturers of single-mode fiber will typically quote four distinct parameters:

The maximum dispersion over a spectral range or "window," e.g.,
$D \leq 3.5$ ps/nm·km from 1285-1330 nm.

The maximum dispersion at a specific wavelength, e.g.,
$D \leq 20$ ps/nm·km at 1550 nm.

The zero-dispersion wavelength with a tolerance, e.g.,
$\lambda_o = 1310 \pm 10$ nm.

The maximum dispersion slope at the zero-dispersion wavelength, e.g.,
$S_o \leq 0.09$ ps/nm^2·km .

Similarly, the bandwidth behavior is a strong function of whether a driving ILD source is operating in a single or multi-longitudinal mode regime (these considerations, not to be confused with fiber modes, will be discussed in Chapter 3).

It is possible to derive an expression, for example, as follows, for the bandwidth, B, for multi-longitudinal mode laser excited single-mode fiber [35]:

$$B \leqslant \frac{1}{8\sqrt{2}[D(\lambda)L\Delta\lambda]^2 + [S(\lambda)L(\Delta\lambda)^2]^2},\qquad(2.2)$$

where,

$D(\lambda)$ is the fiber dispersion at wavelength λ,

$S(\lambda)$ is the slope of the dispersion at wavelength λ,

$\Delta\lambda$ is the rms width of the spectral power density of the source, and

L is the fiber length.

With the manufacturer's data, it is thus possible to calculate the bandwidth of the fiber for a known laser characteristic.

Note from the above expression that the bandwidth of a single-mode fiber varies as $1/L$ when illuminated by a multi-longitudinal mode laser. It can be shown that it varies as $1/\sqrt{L}$ for single longitudinal mode laser excitation [36].

ACCEPTANCE ANGLE

It is of interest to find the angle defining the cone within which light must enter a fiber in order to be guided or *accepted* (see Figure 2-20).

Recalling Snell's Law from Eqn. 2.1, we apply it repetitively, first at the face of the fiber, then at the core-cladding interface.

Thus:

$$n_0 \sin \theta_0 = n_1 \sin \theta_1 \qquad (2.3)$$

and

$$n_1 \sin \theta_1' = n_2 \sin \theta_2 . \qquad (2.4)$$

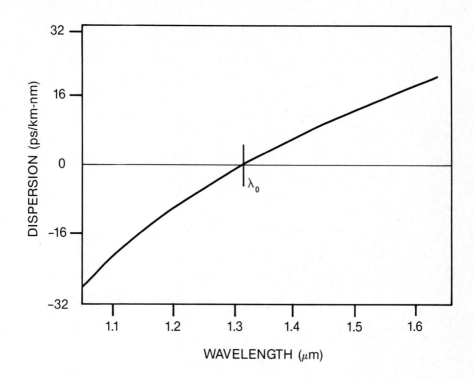

FIG. 2-18. Single-mode fiber dispersion characteristics. (Adapted from [37].)

It will be seen that

$$\theta_1 = \pi/2 - \theta_1' ,$$

so that

$$\sin \theta_1 = \cos \theta_1' . \qquad (2.5)$$

From 2.3 and 2.5,

$$\sin \theta_0 = n_1/n_0 \cos \theta_1' , \qquad (2.6)$$

and from 2.4,

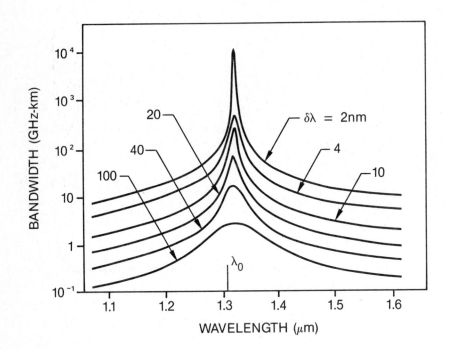

FIG. 2-19. Single-mode fiber bandwidth characteristics.
(Adapted from [37].)

$$\sin^2 \theta_1' = (n_2/n_1 \sin \theta_2)^2 \ . \qquad (2.7)$$

But

$$\cos \theta_1' = \sqrt{1 - \sin^2 \theta_1'} \ , \qquad (2.8)$$

and using 2.7,

$$= \sqrt{1 - [(n_2/n_1) \sin \theta_2]^2} \ .$$

Substituting into 2.6,

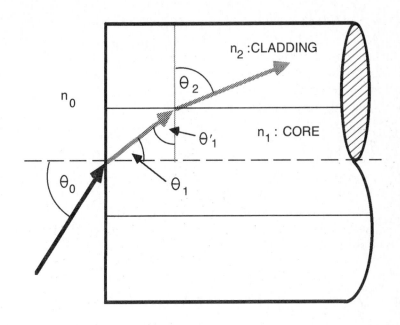

FIG. 2-20. Acceptance angle.

$$\sin \theta_0 = n_1/n_0 \sqrt{1 - [(n_2/n_1) \sin \theta_2]^2} . \qquad (2.9)$$

This is a general expression which, when evaluated for $n_0 = 1$ (free space, or approximately for air), and $\theta_2 = \pi/2$, the condition for total internal reflection, becomes

$$\sin \theta_0 = n_1 \sqrt{1 - (n_2/n_1)^2}$$

$$= \sqrt{n_1^2 - n_2^2} \qquad (2.10)$$

$$\equiv \text{ the Numerical Aperture or (NA) .}$$

Thus, only light within the cone defined by θ_0 can be guided by the fiber structure. θ_0 is therefore termed the *acceptance angle*, and its sine is called the *numerical aperture* (NA). Clearly, the larger the acceptance angle and the numerical aperture, the better the fiber in the sense that a larger proportion of the light offered by a source will be coupled into the fiber for transmission.

Unfortunately, larger NAs usually imply greater dispersion (lower bandwidth). It is easily shown, for example, using geometric optics (see Exercises), that the dispersion of a step-index fiber is proportional to the square of the NA.

Note that the relation defining the numerical aperture lends itself to a geometric interpretation: a right triangle with n_1 for the hypotenuse, and n_2 and NA for the other two sides. Note further that one of the angles of the triangle is the critical angle (for total internal reflection) for the core-cladding interface.

Of course, light rays entering a fiber end are not constrained to be in a plane including the axis (*meridional*). A significant proportion of the launched power will in fact follow paths described as *skewed*.

Consider Figure 2-21, where a skew ray enters the fiber at an angle of θ_s relative to the normal, and at a distance d from the axis. It can be shown that the acceptance angle for such a ray is given by:

$$\sin \theta_s = \frac{\sin \theta_a}{\cos \gamma} , \qquad (2.11)$$

where θ_a is the meridional ray acceptance angle, and γ is one-half the angle between the projections on the fiber face of successive skew rays before and after an internal reflection [38].

Note that skew rays enjoy a larger acceptance angle than do meridional rays, but suffer from a greater number of reflections per unit length. Further, assuming the fiber undergoes no bends, a skew ray will be limited to a cylindrical tube bounded by the core-cladding interface and a *caustic surface* specific to the skew angle and point of entry (see Figure 2-22).

The NA discussed thus far is often referred to as the *theoretical* NA because it differs from *measured* NA. Measured NA as defined by the Electronics Industries Association (EIA) is a far-field measurement which is typically about 8% smaller than the theoretical NA. Care should be taken to be sure which is being referred to.

OTHER UNIQUE PROPERTIES

There are a number of properties that are peculiar to optical fiber, some of which are uniquely advantageous.

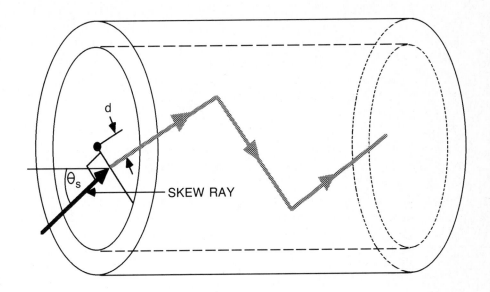

FIG. 2-21. Skew ray acceptance angle.

Microphonics

Because fiber dimensions and curvature affect its ability to guide light, changes in these parameters can modulate the light traversing a fiber.

Consider the arrangement of Figure 2-23. A coil of fiber is attached to a diaphragm capable of translating input sound pressure level to spatial deformation of the coil. The curvature of the coil provides a bias that prevents distortion much as does a biasing magnetic field in a dynamic microphone. The disturbing displacement increases or decreases the curvature, increasing or decreasing the amplitude of the light power that the fiber can convey. The effect is multiplied by the coil because it is manifested upon each turn.

The mass of the fiber coil can be quite small, allowing the frequency response of the structure to be limited by the mass of the diaphragm.

Thus, such a structure might find application in microphones or possibly in phonograph pickups, where the low mass at the pickup and immunity to electromagnetic interference in the leads to the preamplifier are very desirable.

Strain gauges have also been fashioned [39] which capitalize on the same properties.

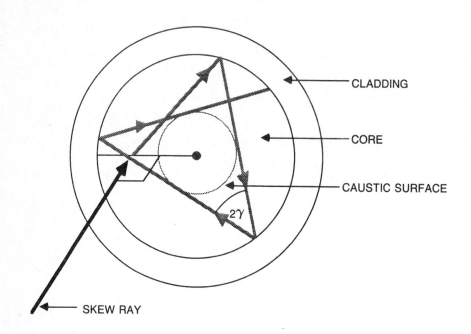

FIG. 2-22. Skew ray trajectory.

Coupled Structures

In single-mode fiber waveguides, fields of significant magnitude exist in the cladding, and if two such structures are placed adjacent to each other, light may be made to couple between them. This property has been applied in a number of novel devices, including switches [40], modulators [41], and sensors [42].

Rotational and acoustic sensors have been fashioned, for example (Figures 2-24 and 2-25), utilizing *directional couplers* comprised of two polarization-maintaining D-shaped fibers brought into intimate contact [43]. (Directional couplers couple light between structures preferentially, depending upon the direction of transmission.)

Figure 2-24 shows a rotation-sensing arrangement suitable for application as a gyroscope element. The signal presented by the laser is split by the

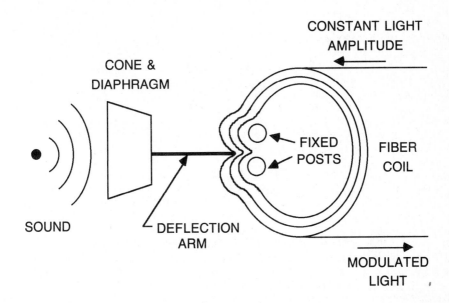

FIG. 2-23. Fiber microphone.

directional coupler into two components, one proceeding clockwise, the other, counter clockwise. The phase difference between the two paths is a measure of the rotation.

Figure 2-25 depicts an acoustic sensor usable as a hydrophone. One coil serves as a reference and the other as the sensor. Again the light from a laser is split via the left-hand directional coupler and traverses the two coils, after which the two beams are recombined by the right-hand directional coupler and conveyed to a phase-difference detector.

2.6 FIBER TYPES

There are two broad classifications of fiber: Multimode and Single-mode. Within the multimode category fall graded-index and step-index fibers.

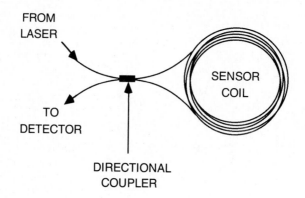

FIG. 2-24. Rotational sensor. (Redrawn with permission; Bulletin 1310, February, 1983, Andrew Corporation).

MULTIMODE - STEP INDEX FIBER

Multimode - step-index fiber has limited application. The dispersion characteristics of such fiber so limit its bandwidth (about 10 - 25 MHz-km) as to make it suitable only for relatively short runs. The major mechanism contributing to the reduced bandwidth is clearly modal dispersion because of the fiber's ability to support many modes.

In practical applications, this fiber type tends to be employed in large core sizes so that coupling power into it is relatively easy.

The most widespread application of such fiber is the distribution of CATV services in large buildings and building complexes. For such applications, a bandwidth of, e.g., 60 MHz, is required for distances of a few hundred feet. A bandwidth-distance product of perhaps 15 MHz-km is thus sufficient, and fiber with a 200 μm core is quite adequate (and has been applied for this purpose [44]).

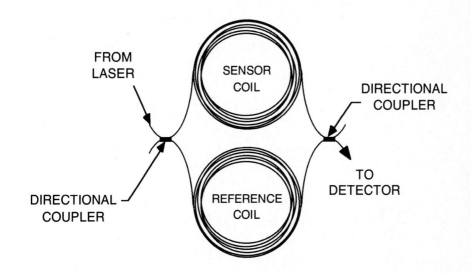

FIG. 2-25 Acoustic sensor. (Redrawn with permission;
Bulletin 1310, February, 1983, Andrew Corporation).

MULTIMODE - GRADED-INDEX FIBER

As described above, graded-index fiber greatly reduces the modal dispersion component of bandwidth limitation by attempting to equalize the flight-time of all ray-paths: the longer physical paths primarily traverse lower index of refraction material so that the axial component of velocity of a nonaxial ray is virtually identical to that of an axial ray. Typical bandwidth is in the 500 - 2,500 MHz-km range.

The index profile is very important to the effective dispersion improvement. The index of refraction in the core may be described by the following expression [45, 46]:

$$n(r) = n_1 \left[1 - 2\Delta \left[\frac{r}{a} \right]^\alpha \right]^{1/2} \tag{2.12}$$

where

$$\Delta = \frac{n_1{}^2 - n_2{}^2}{2n_1^2},$$

n_1 = index of refraction on axis,

n_2 = index of refraction of the cladding,

r = distance from the axis,

a = core radius, and

α = profile parameter (not attentuation).

(Note that $\Delta = \dfrac{(n_1 - n_2)(n_1 + n_2)}{2n_1^2}$ and for $n_1 \approx n_2$, $\Delta \approx \dfrac{n_1 - n_2}{n_1}$.)

A step index profile corresponds to an $\alpha = \infty$. A parabolic index profile is obtained for $\alpha = 2$. The maximum bandwidth is obtained for $\alpha \approx 2$ (the exact value varies with wavelength, and depends on the particular core dopants used).

The act of tube collapse in preform manufacture for certain processes often creates a central "dip" in the profile as an artifact of the former inner surface which tends to be depleted of dopants during the collapse process (see Figure 2-26). Fortunately, its effect is usually minor.

A relatively short time ago, graded-index fiber was the workhorse of the industry, but single-mode fiber now has that distinction. Graded-index fiber now tends to be relegated to specialized short-run applications.

SINGLE-MODE FIBER

Single-mode fiber limits its guidance capability for a given wavelength to one mode via the small diameter of its core and the small core-cladding index difference. It is basically a very small-core step-index structure (see Figure 2-27). In actuality, two orthogonally polarized, strongly-coupled modes are supported in most commercial versions, though special fibers have been produced which are truly single mode [47].

Launching light into a fiber with such a small core (typically less than 10 μm in diameter) is more difficult than with large core fiber, favoring a light source with a highly directional output, and means for co-positioning the source and the fiber axis with precision. Injection laser diodes are nearly ideal as sources for this task, and edge-emitting LEDs can also be used [48]. Surface-emitting LEDs are less effective as launchers into single-mode fiber, as might be expected from examination of their radiation patterns.

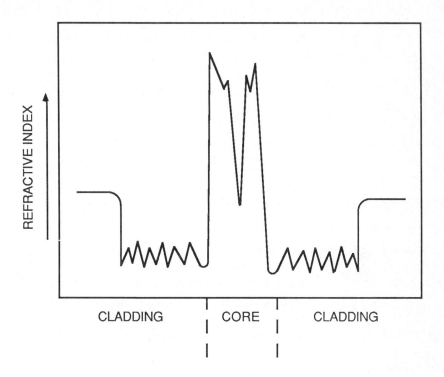

FIG. 2-26. Index profile with central dip.
(Courtesy of R. B. Kummer, AT&T Bell Labs.)

Though graded-index fiber has been widely applied (employed initially, for example, for interoffice trunking in the telephone industry), and some multimode step-index fiber is used in specialized, short-run applications, the future clearly belongs to single-mode fiber.

Single-mode fiber is a variety of step-index structure with a core size and index difference (Δ) chosen such that only the (*hybrid*) HE_{11} mode of propagation is supported at the wavelength of interest. Put another way, the attenuation in such fiber of all modes above the HE_{11} mode is extremely high. (Propagation modes will be considered in detail in Section 2.8.)

The required dimensions of the core of a single-mode fiber as a function of the wavelength of the only allowed propagating mode can be determined from the requirement that

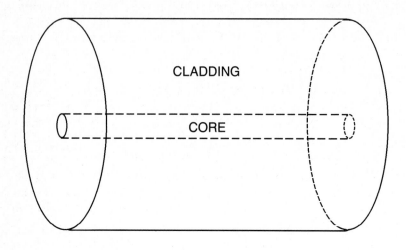

FIG. 2-27. Single-mode fiber.

$$V < 2.405,$$

where V is the *normalized frequency*, a dimensionless parameter given by

$$V = ka\,(n_1{}^2 - n_2{}^2)^{1/2}\,,$$

where $k = \dfrac{2\pi}{\lambda}$ is the propagation constant of plane waves in a vacuum. (The significance of V and the value 2.405 will be better appreciated shortly.)

> Note: Wavelengths are virtually always expressed in terms of their free-space values. This is because their values vary in media of varying indices of refraction. Only the frequency remains constant, and then only if there is no relative motion.

The core radius requirement for single-mode operation can therefore be expressed approximately as

$$a < \frac{1.2\lambda}{\pi(n_1{}^2 - n_2{}^2)^{1/2}}\,.$$

Thus, over the wavelengths of interest, and for typical refractive index choices ($\Delta \approx 0.003$), the core will range from about 8 to 10 μm in diameter. The fiber itself, however, including the cladding, will typically be of dimensions similar to that of graded-index fiber for reasons to be discussed shortly.

It is important to recognize that virtually all single-mode fiber manufactured today is designed to operate in the "slightly multi-mode" regime with a V of the order of 2.8 in order to provide better confinement for the fundamental mode. This is acceptable because the higher-order modes experience very high attenuation (e.g., several dB/m).

The cutoff wavelength for a mode is that wavelength above which a mode becomes leaky and undergoes rapid attenuation. Since the fundamental mode in a single-mode fiber has no cutoff wavelength, that wavelength at which all other modes are at or beyond cutoff is the wavelength at and above which a fiber will be operating in the single-mode regime. This is referred to as the *theoretical cutoff wavelength*. In practice, the *effective cutoff wavelength* is more commonly used. This is the wavelength above which the attenuation of the higher-order modes exceeds some specific value (typically 10 dB/m). The effective cutoff wavelength is generally about 200 nm lower than the theoretical cutoff wavelength.

Standards have been established for the measurement of effective cutoff wavelength of single-mode fiber, e.g., the EIA-455-80 issued by the Electronic Industries Association. Briefly, it specifies (as one technique) that a 2 m length of fiber with a 28 cm diameter loop be measured before and after an additional small diameter loop is impressed to shift the higher-order mode loss edge to shorter wavelengths. That point at which the two measurement curves differ by 0.1 dB is defined as the effective cutoff wavelength.

When a cutoff wavelength λ_c is quoted by a manufacturer, it is an effective cutoff wavelength which has been measured using a method such as that above.

2.7 FIBER SPLICING

Fibers are cabled into moderate lengths for several reasons: e.g., practical manufacturing techniques limit lengths, cable-pulling stresses limit the distances cable can be pulled through conduit, and cable reels must be kept within reasonable size and weight limits, etc. It is necessary, therefore, to employ a means for effectively joining fiber segments. Essentially two techniques are employed, butt coupling and fusing.

BUTT COUPLING

Butt coupling employs the technique of juxtaposing prepared fiber ends in as accurate an alignment as possible, as close as possible, without risking mechanical damage. Losses will occur at the air-glass interfaces of a dry butt splice when the physical alignment is imperfect. Multimode fiber splice losses can be analyzed using geometric optics models and electromagnetic theory, while single-mode systems operate in a regime that is not amenable to conventional geometric optics. However, the single-mode splice can be analyzed as an evaluation of the coupling between "misaligned Gaussian beams" [49] characterized by their *mode field radius*, which is the the radius at which the beam intensity is $1/e^2$ of its value on the axis. (For non-Gaussian beams, there is an alternative definition based on far-field second moments [50]).

Splice loss in single-mode fiber is a function of mode field radius, lateral and transverse offset, angular misalignment, wavelength, and index mismatch (see, for example, Miller [51]).

Index-matching material, typically a gel, an epoxy or a viscous oil approximating the index of refraction of the fiber core, can be used between the fiber ends to reduce losses. Typical losses are 0.05 to 0.5 dB per splice.

The best butt splices utilize a technique known as *active alignment*, which allows mechanical adjustment of the relative positions of the two fiber cores while monitoring the loss [52]. Such structures may in practice have an average loss of less than 0.05 dB for single-mode fiber [53].

FUSION SPLICING

The fiber ends can also be fused together into a continuous structure. A hydrogen-oxygen flame or, more conveniently, an electric arc, can be employed. Small portable units employing microprocessors to align, determine the arc energy, perform the fusion, and test the resulting splice are available [54]. Splices with losses below 0.1 dB are possible with this technique.

Fusion splices are structurally superior to the alternatives and are therefore the technique of choice for applications requiring extraordinary strength such as submarine cable.

2.8 LIGHT GUIDANCE

It will be recalled from the study of metallic waveguides that only a discrete set of modes can propagate in such structures. These modes correspond to the eigenvalue solutions to Maxwell's Equations in view of the boundary conditions imposed by the waveguide inner surfaces. In direct analogy, only a discrete set of propagating modes can be entertained by a fiber.

Simple ray optics provides no clue to the fact that only a discrete set of reflection angles (corresponding to propagation modes) are allowed; indeed, it would appear that a continuum of angles would be permissible. It is therefore convenient to employ wave considerations to understand the limitations imposed. (It is possible, by augmenting ray optics considerations, to obtain the same results when the core dimensions do not approach the wavelength [55].)

THE SLAB WAVEGUIDE

Guidance principles can be illustrated by consideration of a *slab waveguide* comprised of a central layer of refractive index n_1 and thickness h sandwiched between two layers of lower refractive index n_2 (see Figure 2-28).

Consider a lightwave propagating in the z direction via repeated reflections at the interfaces (see Figure 2-29.) For propagation to take place over more than local dimensions, the wavefront after a double reflection must reinforce the wavefront before the first reflection. That is, the net phase shift after two reflections and two inter-reflection transverse-dimension transits must be an integral multiple of 2π.

Therefore,

$$2kn_1h\cos\theta - 4\phi = 2v\pi ,\qquad (2.13)$$

where

k is the (previously defined) free-space propagation constant $= 2\pi/\lambda$,

2ϕ is the phase shift at reflection, and

v is the (integer) *mode number* .

ϕ can be calculated from the Fresnel equation [56] for the TE polarization reflection coefficient R:

$$R = \frac{n_1\cos\theta - \sqrt{n_2^2 - n_1^2\sin^2\theta}}{n_1\cos\theta + \sqrt{n_2^2 - n_1^2\sin^2\theta}} .\qquad (2.14)$$

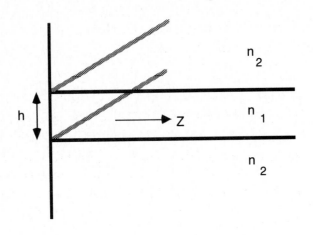

FIG. 2-28. The slab waveguide.

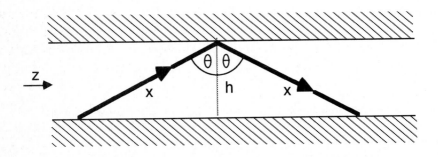

FIG. 2-29. Slab waveguide considerations.

For θ less than the critical angle, R is real [57] and less than unity. For θ greater than the critical angle (total internal reflection), it becomes complex, with unity magnitude, and introduces a phase shift into the reflected wave given by:

$$\phi = \tan^{-1}\frac{\sqrt{n_1^2\sin^2\theta - n_2^2}}{n_1\cos\theta}. \tag{2.15}$$

The phase shift will vary from 0 at the critical angle to $\pi/2$ for $\theta = \pi/2$. Equation (2.13) then becomes

$$\frac{2\pi}{\lambda}n_1h\cos\theta - 2\tan^{-1}\frac{\sqrt{n_1^2\sin^2\theta - n_2^2}}{n_1\cos\theta} = \nu\pi. \tag{2.16}$$

The solution to this transcendental equation for integer values of ν yields the respective allowed incidence angles for propagating modes. Figure 2-30 depicts a set of propagating modes and their corresponding field distributions. Note that the value of ν corresponds to the number of times the field becomes zero transversely within the guiding layer.

Note also that only TE and TM modes can be supported.

More generally, the indices of refraction of the upper and lower bounding media will differ (e.g., glass or Lithium Niobate below and air above as is often the case for Monolithic Integrated Optics [MIO] devices), but that situation complicates the above analysis only slightly.

The phase shift introduced by reflection at the slab interfaces can also be viewed as in Figure 2-31, where the reflection is regarded as having taken place at a boundary layer a distance x beneath the surface of the refractive index discontinuity. The apparent shift in the position of the light ray reemerging from the discontinuity is known as the Goos-Hänchen shift [58].

THE CYLINDRICAL WAVEGUIDE

The situation in step-index cylindrical guiding structures is somewhat more complex in detail but analogous to the slab. Because of the additional complexity of a varying index of refraction in the core, solutions for graded-index fiber are typically found by resorting to approximations.

Analysis may begin with the formulation of Maxwell's equations in a convenient coordinate system, accounting for the peculiarities of the charge- and conductor-free structure, and solving in view of the appropriate boundary conditions. In the step-index case, the permeability, μ, and the permittivity, ϵ, can be assumed to be constant and independent of propagation direction (isotropic conditions) individually in the core and the cladding away from the boundary. (Under anisotropic conditions, these parameters [and the conductivity, σ as well (in general, though it is vanishing in materials of interest)] must be handled as

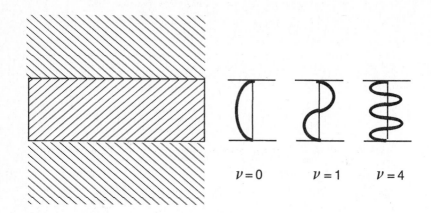

FIG. 2-30. Propagating modes and field distributions.

tensors.)

Consider, then, Maxwell's equations:

$$\nabla \times \mathbf{E} = -\dot{\mathbf{B}} \qquad (2.17)$$

$$\nabla \times \mathbf{H} = \dot{\mathbf{D}} + \mathbf{J} \qquad (2.18)$$

$$\nabla \cdot \mathbf{D} = \rho \qquad (2.19)$$

$$\nabla \cdot \mathbf{B} = 0 , \qquad (2.20)$$

where $\mathbf{E}$ is the electric field intensity,
       $\mathbf{B}$ is the magnetic flux density,
       $\mathbf{H}$ is the magnetic field intensity,
       $\mathbf{D}$ is the electric flux density or displacement,
       $\mathbf{J}$ is the current density, and
       ρ is the charge density.

Because glass is an insulator, $\mathbf{J}$ is effectively zero everywhere, and it may be assumed that there exists no charge in the region of interest, so that ρ is zero.

Maxwell's equations therefore reduce to:

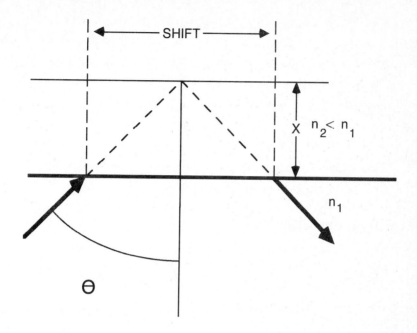

FIG. 2-31. The Goos-Hänchen shift.

$$\nabla \times \mathbf{E} = -\dot{\mathbf{B}} \qquad (2.21)$$

$$\nabla \times \mathbf{H} = \dot{\mathbf{D}} \qquad (2.22)$$

$$\nabla \cdot \mathbf{D} = 0 = \nabla \cdot \mathbf{B}. \qquad (2.23)$$

Note the implications of the last equation: it requires that both magnetic *and* electric lines of force must form closed loops. This requirement places strong constraints upon the variety of propagation modes.

Taking the curl of Eqn. (2.21) yields

$$\nabla \times (\nabla \times \mathbf{E}) = -\nabla \times \dot{\mathbf{B}} . \qquad (2.24)$$

From the identity,

$$\nabla \times (\nabla \times \mathbf{E}) = \nabla (\nabla \cdot \mathbf{E}) - \nabla^2 \mathbf{E} \qquad (2.25)$$

$$= -\nabla^2 \mathbf{E} \text{ in view of (2.23).} \qquad (2.26)$$

Therefore,

$$\nabla^2 \mathbf{E} = \nabla \times \dot{\mathbf{B}} = \frac{\partial}{\partial t}(\nabla \times \mathbf{B}). \qquad (2.27)$$

Using

$$\mathbf{B} = \mu \mathbf{H} , \qquad (2.28)$$

where μ is the permeability,

$$\nabla^2 \mathbf{E} = \frac{\partial}{\partial t}(\nabla \times \mu \mathbf{H}) = \mu \frac{\partial}{\partial t}(\nabla \times \mathbf{H}) . \qquad (2.29)$$

Substituting from (2.22) and using

$$\mathbf{D} = \epsilon \mathbf{E} , \qquad (2.30)$$

where ϵ is the permittivity,

$$\nabla^2 \mathbf{E} = \mu \frac{\partial}{\partial t}\dot{\mathbf{D}} = \mu\epsilon \frac{\partial^2 \mathbf{E}}{\partial t^2} . \qquad (2.31)$$

Similarly,

$$\nabla^2 \mathbf{H} = \mu\epsilon \frac{\partial^2 \mathbf{H}}{\partial t^2} . \qquad (2.32)$$

Both (2.31) and (2.32) represent three distinct equations relating the time and space variations for each coordinate, e.g.,

$$\frac{\partial^2 E_z}{\partial x^2} + \frac{\partial^2 E_z}{\partial y^2} + \frac{\partial^2 E_z}{\partial z^2} = \mu\epsilon \frac{\partial^2 E_z}{\partial t^2} \qquad (2.33)$$

for Cartesian coordinates.

For a sinusoidal excitation, $\mathbf{E}$ can be expressed in a form to display the time and phase dependence:

$$\mathbf{E} = \mathrm{Re}\left[E\left(x,y\right)e^{j\left(\omega t - \beta z\right)}\right].$$

(2.34)

(where β is the *axial propagation constant* or the *axial phase constant* in the core) so that Eqn. (2.33) becomes

$$\frac{\partial^2 E_z}{\partial x^2} + \frac{\partial^2 E_z}{\partial y^2} - \beta^2 E_z = -\omega^2 \mu \epsilon E_z$$

or

$$\frac{\partial^2 E_z}{\partial x^2} + \frac{\partial^2 E_z}{\partial y^2} = \left(\beta^2 - \omega^2 \mu \epsilon\right) E_z.$$

(2.35)

$\omega^2 \mu \epsilon$ is often expressed as $k^2 n^2$, where k $(=\frac{2\pi}{\lambda})$, identified before as the free-space propagation constant, is also known as the *wave number*.

$\sqrt{\beta^2 - \omega^2 \mu \epsilon} = \sqrt{\beta^2 - k^2 n^2}$ is the *transverse propagation constant* in, e.g., the central layer of a slab geometry.

For a cylindrical waveguide, the above expression becomes more interesting:

$$\frac{\partial^2 E_z}{\partial r^2} + \frac{1}{r}\frac{\partial E_z}{\partial r} + \frac{1}{r^2}\frac{\partial^2 E_z}{\partial \theta^2} = -\beta^2_T E_z,$$

(2.36)

where $\beta_T = \sqrt{n_1^2 k^2 - \beta^2}$ is the transverse propagation constant in the core. (Appendix B details the corresponding expressions for H and the other components of E.)

It yields to the process of separation of variables, assuming

$$E_z = A_z\left(r\right) \cdot B_z\left(\theta\right).$$

(2.37)

The (simplified) solutions are:

$$B_z\left(\theta\right) = \begin{cases} \cos\left(\nu\theta\right) \\ \sin\left(\nu\theta\right), \end{cases}$$

(2.38)

where the ν are integers, and

$$A_z(r) = \begin{cases} CJ_\nu(\beta_T r): & \textit{In the Core} \\ DK_\nu(|\beta_T|r): & \textit{In the Cladding}, \end{cases} \qquad (2.39)$$

where J_ν = Bessel Functions of the first kind of order ν,

K_ν = Modified Bessel Functions of the first kind of order ν, and

$|\beta_T| = \sqrt{\beta^2 - n_2^2 k^2}$ is the *transverse decay constant* in the cladding.

(Bessel Functions are discussed in Appendix B.) C and D are constants.

The simplification in the above ignores the constant angle which could be added to Eqn. 2.38, and the Neumann and I functions which are also solutions of Eqn. 2.36, but cannot fit the physically meaningful requirements of the fiber. More specifically, the Neumann functions are infinite at the origin, and I functions diverge at unbounded distances from the origin; both situations are unsatisfactory representations of reality.

The solutions for the axial components may be substituted into the expressions for the radial and azimuthal components to yield expressions for all components subject to boundary conditions.

The boundary conditions which must be satisfied by the solutions require that the tangential fields remain constant across the core-cladding boundary. This means that the z and θ components of both the E and H fields remain constant (explaining the inevitability of an evanescent field in the cladding) while the r components can be discontinuous in proportion to the ratios of the ϵ's and μ's, respectively. Since μ can be expected to be constant throughout the fiber, and, in particular across the core-cladding boundary, of the six E and H components, only the radial component of the E field is discontinuous. The resulting solutions correspond to allowed modes, the first few of which are depicted in Figures 2-32 to 2-35.

The propagation modes are expressed in terms of $TE_{0,\mu}$, $TM_{0,\mu}$, and, for $\nu \geqslant 1$, linear combinations of the two, termed *hybrid modes*, $HE_{\nu,\mu}$ and $EH_{\nu,\mu}$, where ν as before is the order of the Bessel and modified Bessel functions, and is the azimuthal mode number, and μ identifies the μth root of the respective function. (Note: the hybrid designations are as much a matter of custom as of reason [59].)

Figure 2-32 shows both a cross-section and a cut-away of a fiber to better display the E and H fields for the HE_{11} mode. Though this mode is only the lowest order of a multitude of modes supported by multi-mode fiber, it is also the only mode supported by single-mode fiber, and its properties should be remembered in that light. The E field vectors are shown as solid lines, and H field vectors as dashed lines. The HE_{11} mode is a hybrid mode in the sense that both E and H fields may have axial components. Noting that the *Poynting Vector Theorem* demands that propagation occur only in the direction of the E×H vector, and since both E and H fields must describe closed paths, it is not difficult to see that the vectors must describe patterns as indicated. Note particularly that both electric and magnetic field vectors have z components, that they describe closed loops, and the E×H vectors at both the front face and the rear of the depicted cell, point in the +z direction.

Figure 2-33 shows the fields for the TE_{01}, transverse electric field mode. For the electric field to have no axial component and form a closed path, it must describe a planar path normal to the axis as shown.

Figure 2-34 displays the TM_{01}, transverse magnetic field mode which will be seen to be complementary to the TE_{01} mode.

Most higher-order modes are hybrids and are more difficult to depict and visualize. Figure 2-35, for example, shows the electric field for the HE_{21} mode.

It can be shown that a step-index fiber can support about $V^2/2$ propagating modes, and a graded-index fiber with a parabolic profile can support about $V^2/4$ propagating modes [60].

Figure 2-36 shows a dot pattern of light intensity for an actual excited fiber [61]. The number of dots circumferentially is 94, and the number radially is 7. It may be concluded that the mode represented is the $HE_{47,7}$ because circumferential (azimuthal) variation is due to the $\sin \nu\theta/\cos \nu\theta$ term, and displays 2ν maxima in a complete circuit, and the radial (transverse) variation is due to the Bessel Function term, and evidences μ maxima radially.

Though the propagation modes are discrete and finite in number, the nonpropagating modes comprise a continuum. These fall in the categories of *evanescent* modes which decay exponentially in the cladding, and *radiation* cladding modes with imaginary propagation constants.

For the small differences in refractive indices typically employed in fibers, the guidance is "weak" [60], and, in view of the boundary conditions nominally to be satisfied (continuity of azimuthal and longitudinal E and H and radial D fields) at the core-cladding interface, the following approximate eigenvalue equation may be written [62]:

$$U\frac{J_\nu(U)}{J_{\nu\pm1}(U)} = \pm (-W)\frac{K_\nu(W)}{K_{\nu\pm1}(W)}, \qquad (2.40)$$

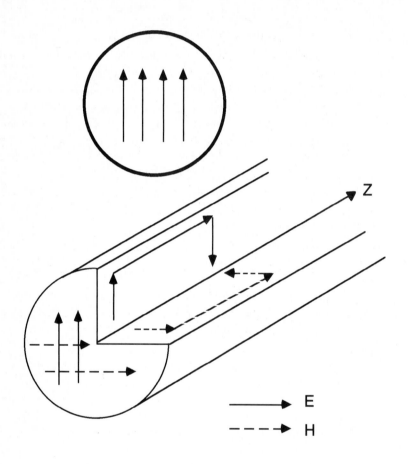

FIG. 2-32. Field distributions for the HE$_{11}$ mode.

where J_ν and K_ν have been defined, the upper signs apply for TM, TE, and EH modes and the lower for the HE modes, and

$$U = \beta_T a \text{ , in the core,} \tag{2.41}$$

$$W = |\beta_T| a \text{, in the cladding, and} \tag{2.42}$$

$$\beta \approx n_2 k. \tag{2.43}$$

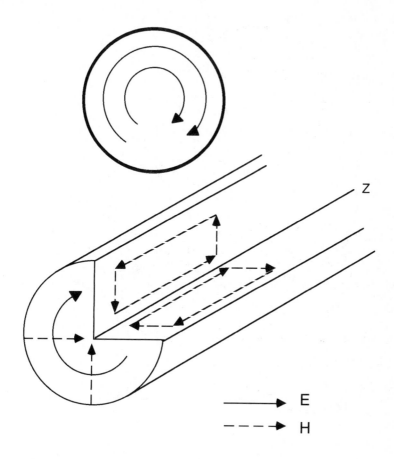

FIG. 2-33. Field distributions for the TE$_{01}$ mode.

U is often called the *normalized transverse phase constant* in the core, and W, the *normalized transverse attenuation constant* in the cladding.

Note that U and W are related to the previously defined V by

$$V^2 = U^2 + W^2 .$$ (2.44)

Cutoff occurs when $\beta = n_2 k$, because then W = 0, corresponding to no attenuation in the cladding; for $\beta < n_2 k$, W becomes imaginary, meaning that propagation of unguided radiating modes is occurring in the cladding.

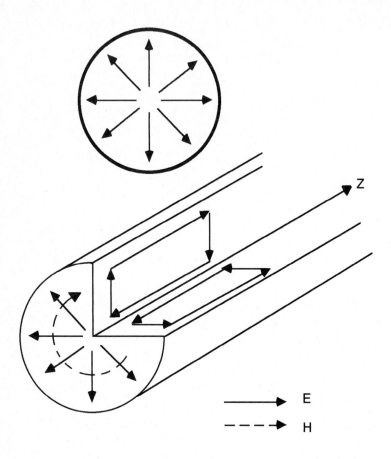

FIG. 2-34. Field distributions for the TM_{01} mode.

From Eqn. (2.40), for $W = 0$ (cutoff), $J_\nu (U) = 0$. But, since $U = V$ for $W = 0$ from Eqn. (2.44), $J_\nu (V) = 0$ also.
(Note in particular that $J_0(2.405) = 0$.) For $W \rightarrow \infty$, $J_{\nu \pm 1}(U) = 0$; this is the far-from-cutoff condition.

It is important to note that though there exists a cutoff frequency for all other modes, the HE_{11} mode has none. Put another way, for a given excitation wavelength, there exists no core diameter which will cause the HE_{11} mode to cease propagating.

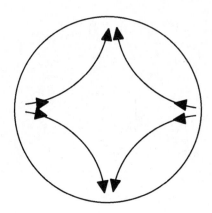

FIG. 2-35. E Field distribution for the HE_{21} mode.

In a sense, there are only two cases of practical interest for transmission: single mode and "many" mode. The latter case applies in large, step-index and graded-index fibers which may support hundreds to thousands of modes at typical wavelengths; aside from the generalized modal dispersion problem, the detailed nature of the multitudes of propagating modes is of little interest.

The analysis for graded-index fiber is additionally complicated by the nonhomogeniety of the refractive index, but it yields to the same general approach, typically employing the Wentzel-Kramers-Brillouin (WKB) [63] method, which is an approximation valid for gradually changing refractive index (i.e., small change in distances of the order of a wavelength).

2.9 OTHER SINGLE-MODE CONSIDERATIONS

It is possible at this point to appreciate other considerations pertinent to single-mode fiber construction.

INDEX PROFILES

Thus far only the simple step-index profile has been discussed in relation to single-mode fiber. Much of the existing practical single-mode fiber employs variations upon the simple step. Figure 2-37 displays a few examples.

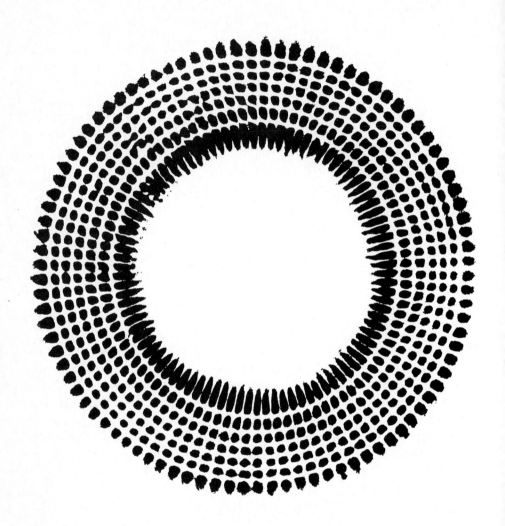

FIG. 2-36. A fiber excited with the $HE_{47,7}$ mode.
(Reproduced with permission from IEE Reprint [61].)

2-37a is the simple step-index structure.

2-37b is a depressed-cladding structure that has the advantage of allowing lighter doping of the core, resulting in lower attenuation fiber, and reduced bending and microbending loss.

2-37c is a triangular profile that has proven particularly effective in producing dispersion-shifted fiber. This structure leads to improved confinement of power, but also generally introduces a finite cutoff wavelength for the fundamental mode [64].

2-37d is a segmented profile structure which is advantageous in producing dual zero-dispersion fiber [65], often referred to as dispersion-flattened or dispersion-broadened fiber.

THE ROLE OF THE CLADDING

The cladding in single-mode fiber is not made thick merely for mechanical reasons. The fields within single-mode fiber extend significant distances into the cladding, and, in fact, a significant proportion (e.g., 20% or more) of the conveyed power traverses the cladding.

In order to maintain low-attenuation and good bending-loss performance, two things are required of the cladding:

It must be of high quality (low loss) material, and

It must be sufficiently thick to shield the guided power from the lossy jacket.

The first requirement is typically met by depositing the cladding in the same manner as the core (e.g., with the MCVD process, via repeated passes). Because producing the high quality cladding is expensive, both in materials and time, its dimensions should be as small as possible consistent with adequate attenuation characteristics. As will be seen, however, the dimensional requirement is demanding.

The cladding employed in single-mode fiber is, then, typically quite thick. A rule of thumb at one time was that the cladding diameter should be seven to ten times the core diameter in order to make the loss due to the jacket negligible [66], but modern designs often reduce this ratio. (As described earlier, in some designs a more complex cladding refractive index profile is employed than the simple, step function type addressed here).

The dependency of jacket-related loss upon cladding thickness is very strong, but so are the costs of uncabled fiber manufacture. In particular, since the volume of material is proportional to the square of the radius of the cladding (neglecting the core), a reduction by a factor of two (were it achievable) in the radius of the cladding would save roughly three-quarters of the material,

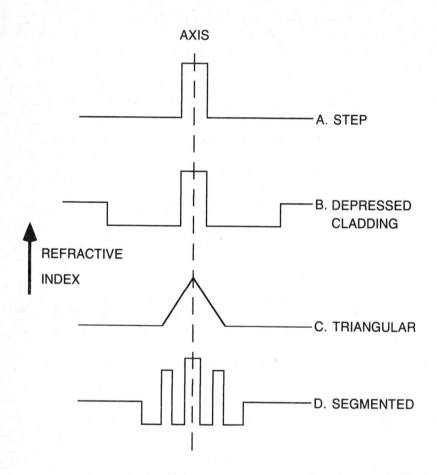

FIG. 2-37. Single-mode fiber index profiles. (Reproduced with permission from Midwinter [65].)

and a corresponding amount of time on, e.g., a glass lathe, to deposit the material. Such savings are diminished somewhat in significance by the fixed costs of the cabling structure and installation, but may be significant enough to substantially alter the overall cost of short link systems, and promote the use of a high-loss variety of single-mode fiber in such applications.

It is possible to formulate the necessary ratio of cladding radius to core radius to achieve a given loss in dB as follows:

$$b/a = \frac{1}{2W}\ln\left\{\frac{8.7W\lambda[1-(W/V)^2]}{2\pi a^2 n}\right\} - \frac{1}{2W}\ln\alpha, \quad (2.45)$$

where

a is the radius of the core,

b is the radius of the cladding,

W is as previously defined,

n is the nominal index of refraction of the cladding,

α is the attenuation (in dB) per unit length, and

V and λ are as previously defined.

(This expression is a modified version of one derived in Miller and Chynoweth [67], which assumes the worst-case attenuation for the jacket.)

(Note that the expression is physically meaningful only for b/a $\geqslant$ 1.)

2.10 COMPARISON: METALLIC VERSUS FIBER WAVEGUIDES

A few words of comparison identifying the analogies, similarities and differences between metallic and fiber waveguides may be useful.

It will be recalled that there exists in metallic slab waveguides a mode (the TEM) which, like the HE_{11} mode for fiber guide, exhibits no cutoff frequency. The TEM mode, however, cannot propagate in a closed metallic (rectangular or circular) waveguide, while the HE_{11} mode can be supported in circular cross-section (e.g., fiber) dielectric waveguides.

Unlike fiber guide, there exists no critical angle in metallic waveguide. An electromagnetic wave will quite happily, for example, reflect ad infinitum transversely at right angles to sides of a rectangular metallic waveguide.

The cutoff of lower order modes occurs at different ratios of fiber core/(circular metallic waveguide) diameter to wavelength in the two transmission media.

The excitation mechanisms of the two guide types are radically different, coupling into metallic guides being accomplished via probes oriented carefully to excite desired modes, while nothing analogous to probes is employed with fiber.

2.11 RADIATION EFFECTS

Three phenomena can result from transient irradiation of optical fiber: luminescence due to ionization-induced and Cerenkov radiation, which can disrupt the signal being transmitted, but is a short-term phenomenon; transient attenuation increase, which may be profound but which is also temporary; permanent attenuation increase, which can be great enough to render a length of fiber unusable [68].

The degree of radiation effect is a function of the core size and the density of impurities. The larger the core diameter, the greater the likelihood of an impinging ray or particle. The greater the density of impurities, the greater the likelihood of an unfavorable-to-transmission local reaction to impingement. Because single-mode fiber enjoys the properties of both lower core cross-section and lower doping levels, it tends to be less vulnerable to radiation damage.

The luminescence effect during irradiation can be profound. A length of only 10 m of low-loss fiber, for example, has been estimated to generate a luminescent output of 1 mW when irradiated at a rate of 10^9 rads per second of 1 MeV photons [69].

The increase in permanent attenuation effects in glass fibers due to X-Ray, gamma ray, electron and neutron irradiation is wavelength dependent, dropping to very low values between 1 and 1.5 μm [70] and rising again for longer wavelengths [71].

Various heat-treating techniques and addition of specialized dopants or dissolved hydrogen to the glass have been tried, but for the most part they have produced only minor improvements or have had side-effects such as increasing the pre-irradiation attenuation and/or irradiation-induced luminescence.

Another approach to radiation hardening is to coat the fiber with aluminum instead of the conventional organic jacketing material. This not only is claimed to increase radiation resistance to 30 krad from a Co 60 source, but additionally hermetically seals against moisture and increases temperature limits to 300 degrees Celsius [72].

Perhaps somewhat surprisingly, plastic fibers are relatively radiation resistant. The expected breaking of long-chain molecules as the result of radiation is not a principle degrading effect; the production of absorption centers is the primary culprit. For short-run applications where radiation resistance is important and other considerations (e.g., bandwidth) do not interfere, plastic fiber may be the medium of choice.

2.12 SUMMARY

Modern optical fiber is primarily manufactured of high-silica glass comprising a central core material and a cladding of lower index of refraction. Light guidance may be explained in terms of the phenomenon of total internal reflection at the core-cladding interface. At least four fiber manufacturing techniques are possible: Outside Vapor Deposition, Modified Chemical Vapor Deposition, Vapor Axial Deposition, and the Double Crucible technique. Fiber cores vary in diameter from less than 10 μm to some 200 μm or more. Fiber attenuation is due to a combination of absorption and scattering. Absorption is due to resonances at the atomic and molecular level, and scattering is due to random structural disorder (Rayleigh scattering), imperfections in materials (Mie scattering), and geometric imperfections (waveguide). Dispersion is due to a wavelength-dependent variation in refractive index, and differential modal group velocities. The numerical aperture is an important parameter associated with a fiber.

There are three basic fiber types, step index (multi-mode), graded index, and single mode. Multimode step-index fiber has a relatively low bandwidth-distance product (e.g., 10 MHz-km), graded-index fiber has an intermediate bandwidth-distance product (e.g., 500 MHz-km - 2 GHz-km), and single-mode fiber has a very large bandwidth-distance product (e.g., 10 GHz-km). Fiber splicing is possible via butt coupling and by fusion. Light guidance can be analyzed simply in terms of a slab waveguide structure, where Fresnel reflection considerations lead to discrete modal transmission. Cylindrical glass waveguides are best understood by analysis paralleling that useful for metallic waveguides, with recognition that the divergence of $\mathbf{E}$ is also zero. Approximations greatly simplify the analysis and yield good results. In particular, the weak-guidance assumption is useful, as is the WKB approximation. Bending robs all modes of power to a greater-or-lesser degree. A variety of index profiles have been applied in single-mode fiber structures, yielding improved characteristics such as dispersion zero shifting and improved macro and microbending loss characteristics. Single-mode fiber generally has somewhat lower attenuation because the core doping can be lower. The cladding in single-mode fiber must be of high quality because 20% or more of the transmitted power traverses the cladding. The major differences (other than cost) between fiber and metallic waveguides are the modal structures (hybrids in fiber), tolerance of a no-cut-off mode (in fibers), and the power launching mechanisms. Radiation in the form of gamma rays, X-rays, and electron or neutron irradiation can temporarily or permanently "darken" fiber.

One particularly telling difference between fiber and metallic alternatives such as coaxial cable, is that while coaxial cable bandwidth can be increased to almost any desired value, a price is paid in the repeater spacing (it must be correspondingly reduced). With fiber, in actual practice, the utilized bandwidth of in-place fiber has been increased via higher performance end electronics and photonics without any reduction in repeater spacing. This is true because the latent bandwidth of fiber is much larger than its usually quoted value; the bandwidth represented by unused additional wavelengths, performance when illuminated by single longitudinal mode sources, and potential use of coherent techniques, is very large indeed.

EXERCISES

1. Derive Snell's Law from basic considerations.

2. Estimate what the cost of materials, labor, and factory overhead must be kept to in order for cabled fiber to be produced at a price of 10 cents per meter. Assume a cost-to-price markup of a factor of two, and a cabling cost equal to the fiber cost.

3. A preform 1 meter in length is to be drawn into 10 km of 100 μm diameter fiber. What must the preform's diameter be?

4. Consider the situation depicted in Figure 2-38, where the light source is embedded within the fiber. Find the effective numerical aperture seen by the light source. Explain why the results are worse than for an external source.

5. Note that the index of refraction of high-silica glass is typically very close to $\sqrt{2}$. If a step-index fiber with a $\sqrt{2}$ core index of refraction is to be manufactured with the highest possible numerical aperture, what must the cladding material be made of? How practical would such a fiber be?

6. Using the dispersion slope characteristics of Figure 2-18, estimate the bandwidth-length product for a fiber operated at its dispersion zero and illuminated by a source with a spectral width of 2 nm.

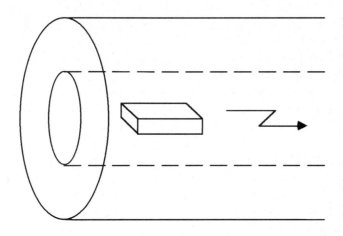

FIG. 2-38. Figure for Exercise 4.

7. Employing geometric optics, show that the modal component of dispersion delay, δ, in a step-index fiber is proportional to the square of the numerical aperture.

8. Find the necessary diameter of the core of a fiber to render it a single-mode structure for a wavelength of 1 μm, $n_1 = 1.4$, and $\Delta = 0.0014$.

9. Consider the following proposition: the cladding used in single-mode fiber is very much thicker than the core, and though of constant index (in the simplest form), it must be of high quality and is therefore expensive. Suppose that, in the interests of reducing costs to the minimum to promote the use of SM fiber for short (e.g., <10 km) spans, the cladding is to be reduced to a point which would increase the attenuation from 0.25 dB/km to 5 dB/km. How much glass (percentage-wise) would be saved?

10. Certain halide glasses are being investigated which are described as having potential for unrepeatered transmission over a distance of 1,000 miles [73].

    a. Making assumptions about launched power and receiver sensitivity, calculate what the attenuation per km of such fiber would have to be.

b. If splices were made every 10 km, how low would the per-splice loss have to be in order to be considered negligible?

c. By way of comparison, note that unamplified voice transmission was (barely) achieved from New York to Denver early in the century. Compute what the attenuation of the copper pair would have had to be in view of the fact that voice becomes unintelligible at about −70 dBm.

11. Check Equation 2.12 for discontinuity of refractive index at the core-cladding boundary. What would be the likely effects of a discontinuity of either sense at this point?

12. Find approximate solutions for Equation 2.16 for the first few modes of propagation.

13. About how many modes will a 50 μm core, step-index fiber with the properties of Exercise 8 support when illuminated by a 1 μm source?

14. Verify Equation 2.44.

REFERENCES

[1] M. V. Klein, *Optics*, John Wiley and Sons, 1970, p. 534.

[2] H. W. Bode, *Network Analysis and Feedback Amplifier Design*, Bell Laboratories Series, 1945.

[3] J. Gowar, *Optical Communication Systems*, Prentice-Hall, 1984, pp. 41ff.

[4] D. N. Payne and W. A. Gambling, "New Low-Loss Liquid-Core Fibre Waveguide," *Electronics Letters*, vol. 8, 1972, pp. 374-376.

[5] D. W. Berriman, "A Lens or Light-Guide Using Connectively Distorted Thermal Gradients in Gases," *Bell System Technical Journal*, vol. 43, no. 4, July, 1964, pp. 1469-1475; and "Growth of Oscillations of a Ray about the Irregularly Wavy Axis of a Lens Light Guide," vol. 44, no. 9, November, 1965, pp. 2117-2132.

[6] F. P. Kapron, *Spectrum,* March, 1985, p. 70.

[7] J. E. Midwinter, *Optical Fibers for Transmission*, John Wiley & Sons, Inc., 1979, p. 188.

[8]   D. B. Keck, P. C. Schultz, and F. Zimar, *U. S. Patent No. 3,737,292.*

[9]   *Optical Fibre Communication,* (CSELT Staff), McGraw-Hill Book Company, 1981, p. 341.

[10] P. J. Lemaire and A. Tomita, "Behavior of Single Mode Fibers Exposed to Hydrogen," *Proceedings of the Tenth European Conference on Optical Communication,* Stuttgart, September 3-6, 1984.

[11] G. W. Tasker and W. G. French, "Low-Loss Optical Waveguides with Pure Fused Silicon Dioxide Cores," *Proceedings of the IEEE,* vol. 62, 1974, p. 1281.

[12] J. B. McChesney, "Materials and Processes for Preform Fabrication - Modified Chemical Vapor Deposition and Plasma Chemical Vapor Deposition," *Proceedings of the IEEE,* vol. 68, no. 10, October, 1980, pp. 1182-1183.

[13] B. Bendow and S. S. Mitra, *Fiber Optics,* Plenum press, 1979, pp. 15-17.

[14] *Fiber Optics & Communication Weekly News,* December 12, 1984.

[15] T. Inoue, K. Koizumi, and Y. Ikeda, "Low-Loss Light-Focusing Fibres Manufactured by a Continuous Process," *Proceedings of the IEE,* vol. 123, no. 6, June, 1976, pp. 577-580.

[16] T. Akamatsu, "Continuous Fabrication of a Phosphate Glass Fiber," *Journal of Lightwave Technology,* vol. LT-1, no. 4, December, 1983, pp. 580-584.

[17] F. T. Dezelsky, R. B. Sprow, and F. J. Topolski, "Lightguide Packaging," *The Western Electric Engineer,* Winter, 1980, pp. 81-85.

[18] J. A. Jefferies and R. J. Klaiber, "Lightguide Theory and Its Implications In Manufacturing," *The Western Electric Engineer,* Winter, 1980, p. 19.

[19] D. B. Keck and Bhagavatula, "Single Mode Fiber Design," *Proceedings of the Sixth Topical Meeting on Optical Fiber Communications,* New Orleans, February, 1984, Paper MF1.

[20] E.g., L. G. Van Uitert and S. H. Wemple, "Zinc Chloride Glass: A Potential Ultralow-loss Optical Fiber Material," *Applied Physical*

Letters, vol. 33, no. 1, 1978, pp. 57-59.

[21] K. Chida, F. Hanawa, and M. Nakahara, "Fabrication of OH-Free Multimode Fiber by Vapor Phase Axial Deposition," *IEEE Journal on Quantum Electronics,* November, 1982, pp. 1883-89.

[22] T. Moriyama, O. Fukuda, K. Sanada, K. Inada, T. Edahiro, and K. Chida, "Ultimately Low OH Content VAD Optical Fibers," *Electronics Letters*, vol. 16, August, 1980, pp. 699-700.

[23] T. Myashita, T. Miya, and M. Nakahara, "An Ultimate Low Loss Single Mode Fiber at 1.55 Microns," *Optical Fiber Communications Conference Proceedings*, Washington, D.C., March 6-8, 1979.

[24] Jefferies and Klaiber, p. 21.

[25] S. E. Miller and A. G. Chynoweth, *Optical Fiber Telecommunications*, Academic Press, Inc., 1979, p. 21.

[26] L. G. Cohen, D. Marcuse and W. L. Mammel, "Radiating Leaky-Mode Losses in Single-Mode Lightguides with Depressed Index Claddings," *IEEE Journal of Quantum Electronics*, vol. QE-18, no. 10, October, 1982, pp. 1467-1472.

[27] P. F. Glodis et al., "Bending Loss Resistance in Single-Mode Fiber," *OFC/IOOC'87 Technical Digest*, TUA3, p.41.

[28] C. M. Miller, *Optical Fiber Splices and Connectors*, Marcel Dekker, Inc., 1986, pp. 86-87.

[29] D. Marcuse, unpublished results.

[30] P. F. Glodis, W. T. Anderson, and J. S. Nobles, "Control of Zero Chromatic Dispersion Wavelength in Fluorine-Doped Single-Mode Optical Fibers," *1983 Optical Fiber Communication Meeting, Optical Society of America*, IEEE #83CH1850-7, p. 12.

[31] V. A. Bhagavatula, M. S. Spatz, W. F. Love, and D. B. Keck, "Segmented-Core Single-Mode Fibres with Low-Loss and Low Dispersion," *Electronics Letters*, vol. 19, 1983, pp. 317-318.

[32] T. Okoshi, *Optical Fibers*, Academic Press, 1982, p. 72.

[33] K. Koixumi, Y. Ikeda, I. Kitano, M. Furukawa, and T. Sumimoto, "New Light-Focusing Fibers Made by a Continuous Process, *Applied*

Optics, vol. 13, no. 2, February, 1974, pp. 255-260.

[34] K. Davies, "Ionospheric Radio Propagation," *Monograph 80*, National Bureau of Standards, Washington, D.C., April, 1965.

[35] L. B. Jeunhomme, *Single-Mode Fiber Optics, Principles and Applications*, Marcel Dekker, Inc. 1983, pp. 129-132.

[36] Ibid.

[37] D. Kalish and L. G. Cohen, "Single-Mode Fiber: From Research and Development to Manufacturing," *AT&T Technical Journal*, January/February, 1987, vol. 66, Issue 1, p. 21.

[38] J. M. Senior, *Optical Fiber Communications Principles and Practices*, Prentice-Hall, 1985, p. 19-22.

[39] J. P. Dakin, "Optical Fiber Sensors - Principles and Applications," *Proceedings of the SPIE - The International Society for Optical Engineering*, vol. 374, 1983, pp. 172-182.

[40] R. V. Schmidt and R. C. Alferness, "Directional Couplers, Switches, Modulators, and Filters Using Alternating Delta-Beta Techniques," *IEEE Transactions on Circuits and Systems*, vol. CAS-26, no. 12, December, 1979, pp. 1099-1108.

[41] K. Tada, and K. Mirose, "A New Light Modulator using Perturbations of Synchronism Between Two Coupled Guides," *Applied Physics Letters*, vol. 25, November, 1974, pp. 561-562.

[42] W. H. Quick, K. A. James, and J. E. Coken, "Fiber Optics Sensing Techniques," *First International Conference on Optical Fiber Sensors*, April, 1983, pp. 6-8.

[43] Optical Fiber Sensor Subsystems, *Bulletin 1310*, Andrew Corporation, February, 1983.

[44] M. F. Mesiya, G. E. Miller, D. A. Pinnowe, "Mini-Hub Addressable Distribution System for Hi-Rise Application," *Technical Papers: Cable'82, NCTA 31st Annual Convention & Exposition*, May, 1982, pp. 37-42.

[45] *Optical Fiber Communication*, p. 16.

[46]   L. M. Boggs and M. J. Buckler, "Testing Lightguide Fiber," *Western Electric Engineer*, vol. XXIV, no. 1, Winter, 1980.

[47]   J. E. Midwinter, "Optical Fibre Communications, Present and Future," *The Clifford Paterson Lecture, Proceedings of the Royal Society: London*, A 392, 1984, pp. 247-277.

[48]   M. Ettenberg, H. Kressel, and J. P. Wittke, "Very High Radiance Edge-Emitting LED," *IEEE Journal on Quantum Electronics*, June, 1976, p. 360.

[49]   C. M. Miller, pp. 139ff.

[50]   W. T. Anderson et al., "Mode-Field Diameter Measurements for Single-Mode Fibers with Non-Gaussian Field Profiles," *Journal of Lightwave Technology*, vol. LT-5, no. 2, February, 1987, pp. 211-217.

[51]   C. M. Miller, p. 143.

[52]   C. M. Miller, pp. 250ff.

[53]   C. M. Miller and G. F. DeVeau, "Simple High-Performance Mechanical Splice for Single-Mode Fibers," *Technical Digest: Conference on Optical Fiber Communication*, San Diego, February 11, 1985, Paper MI2, p. 26.

[54]   T. Tanifuji and Y. Kato, "Realization of a Low-Loss Splice for Single-Mode Fibers in the Field Using an Automatic ARC-Fusion Splicing Machine," *1983 Optical Fiber Communication Meeting, Optical Society of America*, IEEE # 83CH1858-7, p. 14.

[55]   E.g., A. H. Cherin, *An Introduction to Optical Fibers*, McGraw-Hill, 1983, pp. 52ff.

[56]   T. Tamir, *Topics in Applied Physics, Vol. 7: Integrated Optics*, Springer-Verlag, 2nd Edition, 1979, p. 16ff.

[57]   M. Born and E. Wolf, *Principles of Optics*, 4th ed., Pergamon Press, 1970, pp. 110ff.

[58]   T. Tamir, *Integrated Optics*, pp. 25-29.

[59]   T. Okoshi, *Optical Fibers*, Academic Press, 1982, p. 59.

[60]   D. Gloge, "Weakly Guiding Fibers," *Applied Optics*, vol. 10, 1971, pp. 2252-2258.

[61]   P. J. B. Clarricoats, ed., *IEE Reprint Series 1: Optical Fibre Waveguides*, Peter Peregrinus, Ltd., 1975, Cover Page.

[62]   A. W. Snyder, "Asymptotic Expressions for Eigenfunctions and Eigenvalues of a Dielectric or Optical Waveguide," *Transactions, IEEE Microwave Theory Tech.*, MTT-17, 1969, pp. 1130-1138.

[63]   P. M. Morse and H. Feshback, *Methods of Theoretical Physics*, McGraw-Hill, 1953, p. 1092.

[64]   Jeunhomme, p. 47.

[65]   J. E. Midwinter, *The Clifford Patterson Lecture*, pp. 253-4.

[66]   S. E. Miller, pp. 153ff.

[67]   Ibid, p. 153.

[68]   J. E. Gover and J. R. Srour, *Basic Radiation Effects in Nuclear Power Electronics Technology*, Sandia Report, SAND85-0776, May, 1985.

[69]   P. L. Mattern et al., "Effects of Radiation on Absorption and Luminescence of Fiber Optic Waveguides and Materials," *IEEE Transactions on Nuclear Science*, vol. NS-21, no. 6, 1974, pp. 81-95.

[70]   C. E. Barnes, *Radiation Effects in Optoelectronic Devices*, Sandia Report, SAND76-0726, p. 126.

[71]   E. J. Freibele, et al., "Radiation Damage in Single-Mode Optical Fibre Waveguides," *Technical Digest: Conference on Optical Fiber Communications*, Washington, D.C., 1982, pp. 1-9.

[72]   Hughes Aircraft Co., *Photonics Spectra*, February, 1986, p. 88.

[73]   Tasker and French.

CHAPTER 3

ELECTRO-OPTICAL
CONVERSION DEVICES

ELECTRO-OPTICAL CONVERSION DEVICES

Devices capable of transduction of power from the the electrical domain to the light domain range from the incandescent light-bulb to the laser. Two devices are of primary interest at present to the task of economical transmission of signals via fiber optics: the Light Emitting Diode (LED), and the Injection Laser Diode (ILD). Figure 3-1 indicates the spectral properties of the light available from semiconductor sources. Note that an LED generates a much broader spectral output than an ILD, making LED-generated light pulses more sensitive to chromatic dispersion than those issuing from an ILD. For extreme transmission spans, extraordinarily spectrally pure sources are needed, and more refined types of ILDs have been developed for such applications.

Similarly, the power output from an ILD is typically 10 or more times greater than that available from an LED, and a greater proportion of it can be coupled into a fiber because of the small size of the source and the relatively narrow beam width it affords. (One type of LED, known as an edge-emitter, shares the small source-size property of ILDs.)

Devices capable of the reverse transduction (from light power to electrical power) range from photocells to reverse-biased diodes. Again, two devices are of primary interest: the p-i-n diode and the Avalanche Photo Diode (APD). Phototransistors are of some interest, but they tend to have slower response characteristics than the two diode types. Simple reverse-biased diodes are also capable of transduction but are slower and generally less efficient than p-i-ns.

3.1 ELECTRICAL TO OPTICAL CONVERSION

Ideally, a transducer will perform the conversion from one form of energy to another with 100% efficiency. Such ideals regrettably still evade the user of fiber-optics light-launching devices as it has the makers of light-bulbs. Incandescent bulbs, for example, are only about 5% efficient in converting electrical energy into visible light, and fluorescent tubes, 20% efficient. The light-emitting

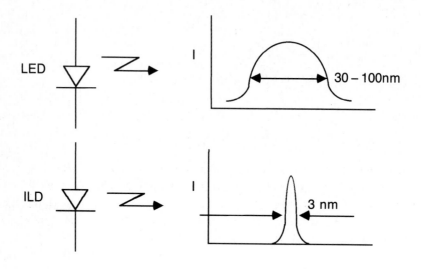

FIG. 3-1. Semiconductor light sources.

devices available for transmission purposes are generally considerably superior to these prosaic devices in internal quantum efficiency (of the order of 50% and higher). Unfortunately, however, the proportion of light that can in practice be launched into a fiber is often much smaller.

Light emission is principally a consequence of a change of the energy state of electrons, the frequency of the light emitted being determined by:

$$W = h\nu,$$

where

W is the energy change in Joules,

h is Planck's Constant $(6.626 \times 10^{-34}$ J/Hz) , and

ν is the light frequency.

Most light emission is the result of random electron transitions, and indeed, such transitions are characteristic of LEDs and of ILDs below the *lasing threshold.* Above threshold, ILDs exhibit ordered, *stimulated emission.* A unique property of the light emitted by electrons stimulated in this fashion is *coherence* (to be defined shortly). There are also operational advantages associated with the rapid response of the light output to variations in the applied electrical signal amplitude.

Radio frequency photons have energies some tens of thousands of times smaller than that of photons at wavelengths of interest for fiber optic systems. Radio frequency photons are created by perturbations in the spin of electrons rather than orbital changes.

The human eye is sensitive in the range of about 0.4 to 0.8 μm; ultra-violet light spans 0.01 to 0.4 μm, and near infrared is from 0.8 to 1.5 μm. Practical optical communication sources operate principally in the near infrared, though some systems are operated in the visible range to accommodate the spectral attenuation properties of plastic media.

HISTORY OF ELECTROLUMINESCENCE

Electroluminescense (the generation of light as the result of current passing through a material) was discovered in 1907 [1], but the first remarkable application of the principle took place in 1954, when Townes did his Nobel-Prize winning work on the MASER (Microwave Amplification through Stimulated Emission of Radiation). This activity was carried out using gaseous (ammonia) and solid state (ruby) working materials [2].

In the 1958-60 period, ruby LASERs (Light Amplification through Stimulated Emission of Radiation) were demonstrated [3].

In 1961-62, GaAs semiconductor *homojunction* (utilizing the same semiconductor material throughout) lasers were first made [4].

In 1962, ternary (GaAsP) lasers were demonstrated [5].

In 1963, single *heterostructure* (utilizing differing semiconductor materials) lasers were constructed [6].

In 1970, the first *double heterojunction* (utilizing three distinct semiconductor material layers) laser (the standard present-day structure) was made, allowing continuous room-temperature operation [7].

INJECTION LASER DIODES

Injection laser diodes convert energy from electrical to optical form via injecting electrons across a p-n junction into an optically tuned cavity, where they occupy an elevated energy state (the *conduction band*). Light stimulation can then induce them to revert to a lower state (the *valence band*), emitting photons in the process.

The conduction band is a band of energy states above a forbidden region separating it from the valence band, which is the highest stable energy band. The energy difference between the lowest subband of the conduction band and the highest subband of the valence band is known as the bandgap energy E_g. In solids, these bands encompass a range of energies, unlike the case in gases, where strict energy levels exist.

Lasing

The ability of certain solid-state substances and gases (e.g., Ruby, HeNe, CO_2, etc.) to support stimulated emission has been known for some time, and devices utilizing this capability have found wide application (e.g., retinopathy treatment [8], industrial wire die manufacturing, surveying [9], weapons ranging [10], etc.).

Electrons in an elevated energy state will, in the absence of stimulation, tend to return to a ground state on a random basis, emitting photons at a wavelength corresponding to the energy difference ($\lambda \approx 1.24/E_g$, where λ is in μm when E_g is in electron volts); see Figure 3-2. When a photon of a wavelength approximately corresponding to the energy state difference impinges upon an excited electron, it stimulates its transition and is joined by the emitted photon. These photons, in turn, if they impinge upon excited electrons, will continue the process, in effect amplifying the signal represented by the original photon.

Nonradiative transitions occur when an excited electron drops to a *trapping state* and eventually to a ground state, converting its potential energy to *phonons* (lattice vibrations), and ultimately, heat.

Stimulated emission will also take place in association with light produced by ordinary sources. However, the vast majority of the light is emitted via random emission in such cases. Special conditions must be met in order for stimulated emission to preponderate.

In a sense similar to the concept of critical mass for nuclear reactions, the density of available excited electrons must be above a critical threshold in order to support lasing action. In ILDs, the threshold translates to a critical current density.

It is important that most of the injected carriers do not succumb to nonradiative recombination, but participate in the useful generation of light. The effective carrier lifetime τ can be defined by [11]

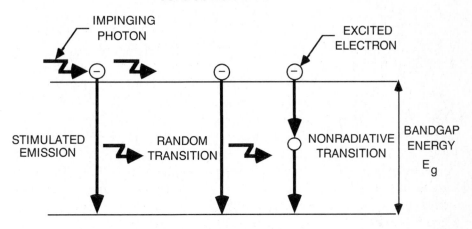

FIG. 3-2. Emission mechanisms.

$$\frac{1}{\tau} \approx \frac{1}{\tau_r} + \frac{2S}{d} + \frac{1}{\tau_{nr}},$$

where τ_r is the radiative carrier lifetime in the active region, S is the nonradiative interface recombination velocity at the interfaces a distance d apart, and τ_{nr} is the nonradiative recombination lifetime due to bulk imperfections.

If τ_{nr} is large relative to τ_r (which is usually the case), then the internal quantum efficiency can be expressed as

$$\eta_i = \frac{\tau}{\tau_r} \approx \frac{1}{1 + \frac{2S\tau_r}{d}}.$$

Coherence

The impinging photon and the stimulated photon are of the same wavelength and propagate in the same direction with a constant phase relationship; such photons are said to be *coherent*.

Structures

Several structures have been employed in the fabrication of ILDs, with interest growing in spectrally pure sources. The basic ILD is a *Fabry-Perot* resonator which, though effective as a transducer, and much more spectrally pure than an LED, will, in general, entertain a multiplicity of *longitudinal modes* spread over perhaps 5 to 10 nm in wavelength, even when operated unmodulated. Direct modulation, because it varies the carrier concentration in the laser, also alters the index of refraction and therefore the wavelength, so that wavelength "chirping" typically occurs at the leading and trailing edges of a modulating pulse. Chirping is also a function of the rate of change of the light level.

Techniques which have been employed to preserve spectral purity (single longitudinal mode operation) include incorporation of internal or external periodic structures, separate active or passive tuning cavities, external mirrors, injection locking with external sources, and indirect modulation.

● *The Fabry-Perot Structure*

A Fabry-Perot structure consists of a cavity, with mirrors at either end, which functions as a resonator analogous to an organ pipe (see Figure 3-3). One or both of the mirrors is only partially reflective so that useful light can be coupled out. Typically, the mirrors are effected via cleaved semiconductor-air interfaces, though coatings can also be utilized to adjust reflectivity at one or both interfaces.

The other four sides of the cavity are typically designed to confine the light emitted within the cavity via guidance with abrupt refractive index changes.

Early homostructure devices suffered from the need for extremely high currents to bring about lasing due to leakage of charge and light from the active region, making them operable only at low temperatures; the latest, and most widely employed, are the double-heterostructure devices, which have manageably low current requirements, allowing room-temperature operation.

Heterostructure devices are comprised of a layered arrangement of carefully chosen alloy materials. The layering is performed via crystal growth, and the material choice is critical because it must accomplish three goals: provide an appropriate bandgap difference to confine minority carriers to the active region; establish an adequate difference in refractive index to confine light; and effect a

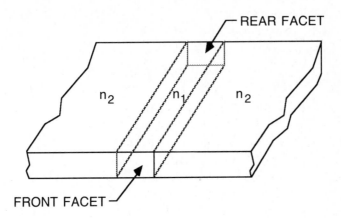

FIG. 3-3. The Fabry-Perot laser structure.

sufficiently close crystallographic match such that few defects result at the interface. (Crystal defects have been identified as one of the major causes of shortened life [12].)

Heterostructures are beginning to be employed in the fabrication of high-performance bipolar transistors as well [13]. Briefly, the advantages potentially conferred upon bipolar transistors via utilization of heterojunction technology considerably surpass those available to FETs, and 100 GHz oscillation frequencies and 10 ps switching speeds have been predicted.

The function of the Fabry-Perot resonator relative to lasing can be better understood by examining the process which occurs within its confines. At equilibrium, energy is flowing into and out of the structure at equal rates. The incoming energy is electrical and is converted to heat and useful light. The material may be viewed as optically lossy, attenuating the light produced via absorption and leakage from the active region as a function of the distance traversed (which may be considerable because of the repetitive reflections at the device's ends). Loss (of a useful nature, since light must be allowed to issue from the device) also occurs at the cleaved facets, which, in general, allow a portion of the impinging light to pass through, and reflect the remainder; the reflectances of the two facets may be independently chosen. Thus, for example, the rear facet may be made to have near 100% reflectance if there is no utility to monitoring the output from that port. The front facet must, however, have finite transmittance to allow useful light output.

The losses must be compensated by *gain* in the structure, and the value that parameter must attain can be calculated from the following (see Figure 3-4):

Let the two facets be separated by L meters and have reflectances of R_1 and R_2. Let the loss per unit distance be given by α, and the gain per unit distance be given by g (both in units of nepers/meter). Consider a packet of light in the center of the structure which is propagating to the left; for sustained operation, the power of the light at that point must remain the same after traversing half the cavity to the left, being reflected at the back facet, traversing the entire cavity to the right, being reflected at the front facet, then returning to its starting point.

If the power level for the packet at the center is P, then it will have decayed to $P - \alpha L/2 - \ln\dfrac{1}{R_1}$ after it reaches the left end. At the right end, an additional loss of $\alpha L + \ln\dfrac{1}{R_2}$ will have occurred. Returning to the center, another penalty of $\alpha L/2$ will have been exacted. To compensate for this loss, a distributed gain, g, must exist which can be calculated from

$$P - \alpha L/2 - \ln\frac{1}{R_1} - \alpha L - \ln\frac{1}{R_2} - \alpha L/2 + 2Lg = P \ ,$$

or

$$g = \alpha + \frac{1}{2L}\ln\frac{1}{R_1 R_2} \ .$$

Knowledge of the required gain can be used to calculate the current threshold to be surpassed for lasing, since the gain is linearly related to the current density.

Of course, in real devices, neither the charge nor the light are perfectly confined to the cavity, and additional losses are suffered,

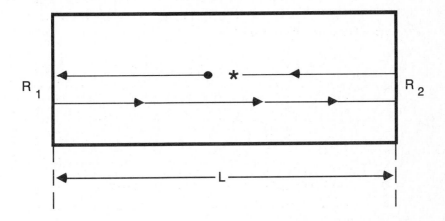

FIG. 3-4. Fabry-Perot gain considerations.

which can be viewed as contributors to an overall α. Nonetheless, the above analysis is instructive relative to the basic mechanisms at work.

A Fabry-Perot semiconductor laser viewed as a light conduit is subject to the same physical laws as those governing fiber, though the indices of refraction are such that guidance is strong, so that weak-guidance approximations are inappropriate. Unlike fiber, however, the axial dimensions of the laser are quite small, and become important in governing the behavior of the device. In particular (to a first approximation), for a given longitudinal mode to be sustained, the length of the device must be an integral number of half wavelengths, i.e.,

$$L = \frac{m\lambda}{2n}, \ m \text{ an integer,}$$

recognizing that λ/n is the wavelength in the device cavity. Several wavelengths may simultaneously satisfy this condition, giving rise to a multiplicity of longitudinal modes and expanding the spectral width of the laser's output. (The spectral width may be specified as full width at half maximum [FWHM], or rms; FWHM ≈ 1.2 rms.)

● *Distributed-Feedback Structures*

A distinctly different device type, the periodic, Distributed-Feedback (DFB) structure [14] has received considerable attention. It differs from the Fabry-Perot structure in that mirrors as such need not be utilized. Rather, a series of regularly spaced discontinuities (e.g., etched lines) is employed whose net effect is essentially that of a frequency-selective mirror. Its lineage is directly traceable to the Bragg modulator [15] which uses acoustically excited pressure waves in water to modulate light.

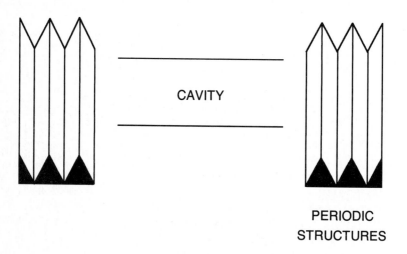

FIG. 3-5. The Bragg-structure laser.

Two advantages accrue from such devices: the lack of a need for mirrored surfaces allows the devices to be fabricated within a crystalline structure rather than abutting air on two sides (this, for example, permits direct coupling of light into integrated optic conduits and devices); the light emitted is more spectrally pure than that of a single-cavity Fabry-Perot diode because only one longitudinal mode is supported.

A variation upon the DFB structure is the Distributed Bragg Reflector (DBR) (see Fig. 3-5), which places the etched gratings near the cavity ends [16].

• *Cleaved, Coupled-Cavity ILDs*

An important variation upon the design of ILDs is the Cleaved, Coupled-Cavity (C^3) ILD [17]. This device can be manufactured from a conventional ILD via cleaving approximately midway along its length (see Figure 3-6).

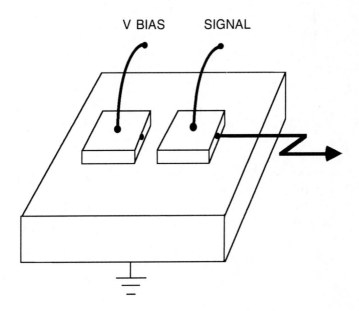

FIG. 3-6. The C^3 laser.

One half of the device is driven in the usual fashion, while the other is provided with a subthreshold bias current. The two devices are free to couple light between the light cavities through the intervening airgap.

C^3 devices have proven capable of single longitudinal-mode operation, but possibly more importantly, they exhibit the capability of being tuned (in discrete steps) to different wavelengths as the bias current supplied to the passive structure is varied [18]. These wavelengths correspond to different single longitudinal modes, and are therefore, very closely-spaced.

The principle may be explained in terms of two independent structures each having a set of longitudinal modes it can support. Ideally, they will share only one of them, which is then mutually supported. Changing the charge-density in the passive structure changes the index of refraction of its cavity material, the set of longitudinal modes it will support, and therefore the mode it will support in common with the active structure, effectively tuning the overall device operating frequency.

Other properties (e.g., bistability [19]) of C^3 devices are still under study, as are variations such as dual modulation of the cavities.

An early C^3 laser was employed in a 160 km (100 mile), 432 Mbps, unrepeatered single-mode fiber demonstration, and a 104 km, 1 Gbps unrepeatered single-mode fiber experiment [20].

- *Other Techniques*

  External mirrors, injection locking and indirect or external modulation are other techniques useful for producing spectrally pure modulated signals. The external mirror augments the cavity's mode selection. Injection locking illuminates the diode with the output of a secondary laser, promoting selection of the incident wavelength. External modulation utilizes a passive modulator following an unmodulated source enjoying a constant carrier density in steady state.

Capabilities and Limitations

ILDs offer power output an order of magnitude or more greater than that of LEDs (typically of the order of 1 to 10 mW, though much higher levels are possible), with better directionality, smaller active spot size (providing better coupling into fiber [as much as 50% for graded index]), and narrower spectral width (see Figure 3-7). Further, they are capable of responding to modulating frequencies far higher than those to which LEDs can typically respond (e.g.,

>> 1 Gbps).

The spectral component spikes of Figure 3-7 correspond to discrete longitudinal modes, each of which may have a linewidth of less than 10^{-5} μm. The collective spectral width of such multi-modal output compared to that of an individual mode explains the interest in single-longitudinal-mode diodes for low-dispersion transmission.

The number, magnitude and spectral position of the modal components are also typically a function of instantaneous drive current and its time rate of change, as the cavity refractive index changes with carrier density.

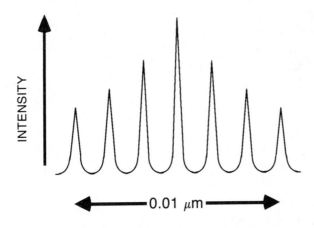

FIG. 3-7. Typical ILD spectrum.

As a class, ILDs suffer from several problems. The lasing threshold current, for example, changes with temperature and age, requiring additional circuitry for monitoring the diode's output and changing the operating point accordingly. ILDs sometimes become unstable and break into very high frequency oscillation. The spectral width of the output of Fabry-Perot types changes with current level during modulation, so that pulse distortion may be evidenced after traversing a length of fiber. The coherence of ILD emission creates problems when a multimode fiber is illuminated; the "speckle pattern" created may shift with fiber temperature and position and materially affect transmission (the *modal noise* effect [21]). One technique for overcoming this problem is superposition of an ultra-high-frequency signal to cause frequent transitions and corresponding "spraying" of spectral components.

The devices are also relatively intolerant to back-reflections from discontinuities, "kinks" in the power output versus current characteristics may appear, and catastrophic facet damage can occur when the devices are overdriven. Further, due to the more demanding structural requirements, ILDs are still considerably more expensive than LEDs.

LIGHT EMITTING DIODES

Light emitting diodes are characterized by their emission of incoherent light due to the random nature of the recombination of the hole-electron pairs. In a sense, they may be viewed as degenerative ILDs because ILDs function as LEDs for current levels below threshold. The spectral width of a typical long-wavelength LED's output is shown in Figure 3-8, where it will be seen that the half-power spectral width is in the neighborhood of 0.1 μm (short-wavelength LEDs typically have a spectral width less than half as great).

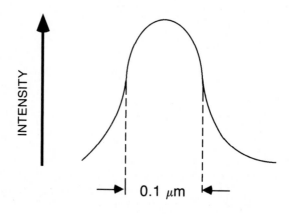

FIG. 3-8. Typical LED spectrum.

Structures

The structure of an LED is, in essence, a simpler form of an ILD device (see Figure 3-9). The two elements generally missing are the mirrors or periodic structures, but the charge and light-confining heterostructures may be present to a greater or lesser degree (as will be seen, edge-emitting LEDs retain much of the structure of an ILD).

Emissions may in general take place in all directions from the light-generating junction area, including edges and unimpeded surfaces. LEDs may be classified on the basis of which of these emission directions is enhanced.

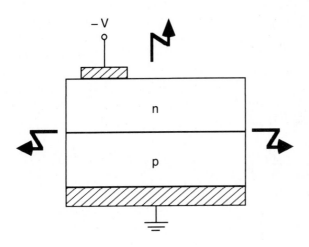

FIG. 3-9. LED structure.

● *Surface Emitters*

Surface emitting devices (e.g., etched-well type Burrus diodes [22]) emit light through a window that is in a plane parallel to the surface of the device (see Figure 3-10). The upper layer is chosen to be nearly transparent at the generated wavelength, the center layer is the active portion, and the large E_g bottom layer confines carriers and guides or reflects the generated light with a relatively low index of refraction. The emitted light is not directional, with a beam width at half intensity of about 120 degrees, and approximately Lambertian properties, and is difficult to couple efficiently into fiber without the aid of lens structures. Spherical lenses are routinely applied in association with these devices (see Figure 3-11). Such

spheres may be comprised of anything from sapphire with an index
of refraction of 1.75, to quartz with an index of 1.52.

Surface emitters exhibit a *radiance saturation* phenomenon at
high current levels due to nonradiative recombination and carrier
leakage (see Figure 3-12) which limits output and contributes to
nonlinearity.

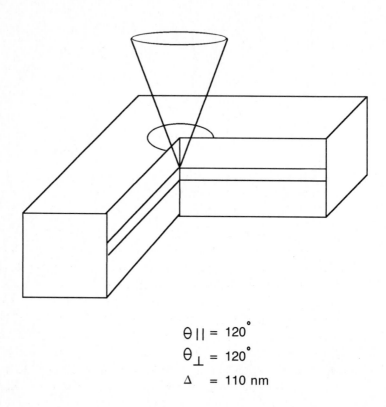

$$\theta \| = 120^{\circ}$$
$$\theta_{\perp} = 120^{\circ}$$
$$\Delta = 110 \text{ nm}$$

FIG. 3-10. Example of a surface-emitting diode.

• *Edge Emitters*

Edge-emitting LEDs emit light through a cleaved section of the
device in a manner similar to that of a Fabry-Perot ILD (see Figure 3-13), except that the radiation pattern is as broad as a

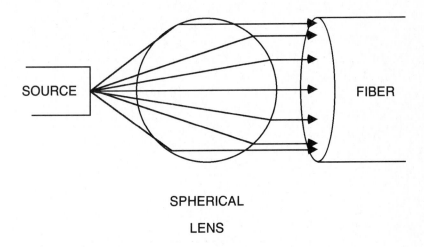

SPHERICAL

LENS

FIG. 3-11. LED-lens combination.

surface-emitting diode in the plane parallel to the junction. These devices typically employ heterojunctions and may provide a measure of ILD-like gain at high current densities; in the latter case, the diodes may display *superlinear* or *superluminescent* behavior (see Figure 3-14) [23]. Some success has been achieved in launching appreciable light into single-mode fiber with these diodes [24], and this capability may boost single-mode fiber into more rapid application. A power level of 250 μW has been coupled into a single-mode fiber from an edge-emitting LED, allowing transmission at 140 Mbps over 80 km [25], and 16 Mbps signals have been transmitted 107 km in a similar experiment [26].

Capabilities and Limitations

Typical output light power of a surface-emitting LED is of the order of 1 mW, but only a few percent of that power can be successfully launched into a suitable (large step-index or graded-index) fiber without lensing, and launching into single-mode fiber is virtually a futile exercise (with the exception of short distance, very low frequency, high receiver sensitivity situations).

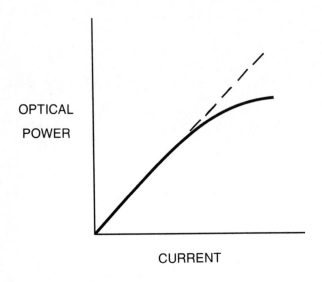

FIG. 3-12. Radiance saturation in surface emitters.

Somewhat more successful coupling is possible from an edge-emitting device as discussed above, but the power output of such devices is generally less than that of surface-emitting types.

Although typical LEDs operate below 100 MHz, 200 MHz is not uncommon, and experimental 2 GHz operation has been reported [27].

LEDs in general, and surface emitters in particular, are significantly less sensitive to temperature than are ILDs because there is no threshold, but their power output for a given current drops about 1% per degree C.

Generally (between devices), speed is inversely proportional to maximum power output because of the higher doping required for higher speed and the correspondingly lower efficiency.

Visible LEDs

LEDs emitting visible light have been in evidence in commercial products for some time. These devices fall in categories associated with the wavelength of their emitted light [28]. Red LEDs are *direct-bandgap*, GaAs substrate devices

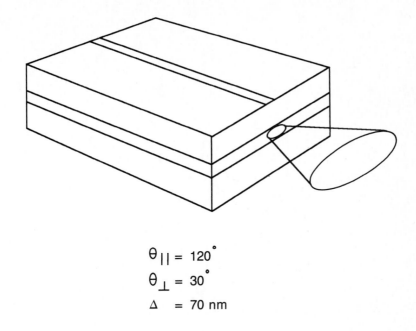

$$\theta_{||} = 120^{\circ}$$
$$\theta_{\perp} = 30^{\circ}$$
$$\Delta = 70 \text{ nm}$$

FIG. 3-13. The edge-emitting LED.

$(E_g \approx 1.8 \text{ eV})$, while orange $(E_g \approx 1.9 \text{ eV})$, yellow $(E_g \approx 2.1 \text{ eV})$ and green $(E_g \approx 2.3 \text{ eV})$ are *indirect-bandgap*, GaP substrate structures (the topic of direct vs. indirect-bandgap materials will be discussed in the next section). Blue LEDs typically produce light via an *up-conversion* process, converting lower energy infrared light to higher energy blue light via the mechanism of multiple "boosts" within a secondary, phosphorescent material. There has also been success in employing silicon carbide for direct production of blue light at about 480 nm $(E_g \approx 2.5 \text{ eV})$ [29].

Aside from the tasks of indicators and displays, visible wavelength emissions are important for illuminating plastic fibers, whose typical (Methyl-Methacrylate) attenuation minimum is at about 0.7 μm.

MATERIALS

It would be very convenient if a material such as silicon, with a wealth of knowledge, experience, and processing equipment behind it, could be employed in the fabrication of light-emitting diodes. Regrettably, however, silicon (and

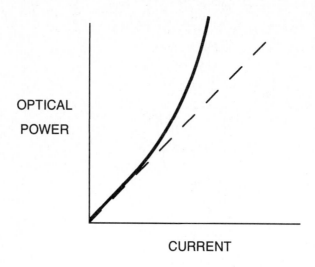

FIG. 3-14. Superlinear/superluminescent edge-emitting LED behavior.

germanium) fall in the category of *indirect bandgap* materials. This means that the transition from an excited state to the ground state is not momentum-preserving because of differing values of wave number (k) at the minimum conduction band energy and the maximum valence band energy [30], and a *phonon* must issue, expending energy. Thus the internal quantum efficiency of a device fashioned of these materials is prohibitively low.

It is therefore necessary to resort to material systems with *direct bandgap* transitions in order to obtain high quantum efficiencies. For short wavelengths (in the range of 0.8 to 0.9 μm), GaAs or AlGaAs may be "tuned" by varying the proportions of the constituents. Similarly, InGaAs or InGaAsP may be employed in the region of 1.2 to 1.4 μm, and InGaAsP/InP in the range of 1.4 to 1.6 μm. The chemical significance of the choice of these III-V compounds may be appreciated by examination of their placement in the periodic table (Figure 3-15). Note that the constituent elements of interest flank the IV column and the compounds will have an average of 4 valence electrons per atom.

(It will be recalled that oxidation is the loss of electrons, and reduction is the gain of electrons. The electron shells progress in the maximum number of electrons as $2n^2$ for n an integer, which places an upper bound on the oxidation number for each element. Rows in the table correspond to elements with the same number of occupied shells, and columns to elements with the same number of electrons in the outer-most shell.)

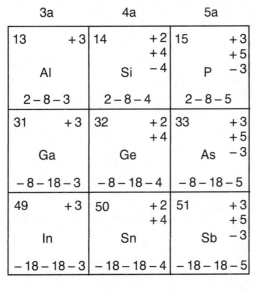

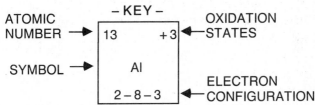

FIG. 3-15. A pertinent portion of the periodic table.

There has been a marked shift of interest from the initially successful short-wavelength devices toward the longer wavelengths because of the reduced fiber attenuation, better material dispersion characteristics, and success in the fabrication of light-launching and receiving devices that can operate at these wavelengths. Short wavelength emitters may eventually become of interest primarily in association with wavelength-division multiplexing applications using short, medium and long wavelengths.

FABRICATION

Several fabrication techniques have been employed in the manufacture of ILDs and LEDs: liquid-phase epitaxy, vapor-phase epitaxy, molecular-beam epitaxy, and recently, metal-organic vapor epitaxy. A technique called vapor-levitation epitaxy has also recently been reported.

Liquid-phase epitaxy brings a substrate sequentially into contact with materials to be deposited which are in a molten state. Epitaxial growth takes place at the solid-liquid interface at a rate that is temperature dependent (temperatures are carefully controlled). Relative motion between the substrate and molten material is necessary as the sequential process progresses. Growth rates can be high.

Vapor-phase epitaxy presents the material to be deposited in a gaseous form. Growth rates are generally less than those possible via liquid-phase processes.

Molecular-beam epitaxy projects a directed beam of molecules to be deposited at the substrate. Though growth rates are low, the purity of the deposited layers is unrivaled. [31]

Metal-organic vapor epitaxy utilizes an organic vehicle to aid in conveying constituents. Reported growth rates are high [32].

Vapor-levitation epitaxy utilizes a method of floating a wafer on a thin layer of reactant gases that stream through a porous quartz disk [33]. The reaction with the wafer promotes epitaxial growth on the wafer bottom.

MULTIWAVELENGTH DEVICES

Multiwavelength devices capable of emitting two or more independently modulated wavelengths are also possible, and have potential advantages with respect to cost and space. The two configurations that have been studied use devices that are in-line or side-by-side [34, 35]. The in-line devices must be fashioned of materials with the property that each successive device is nearly transparent to the emissions of all devices preceding it. The side-by-side devices, being off-axis, are disturbed in their ability to effectively couple light into the fiber.

3.2 OPTICAL TO ELECTRICAL CONVERSION

Conversion from the optical to the electrical domain requires a device which can efficiently collect incident photons and cause them to generate hole-electron pairs which can in turn be detected electrically. Other ideal features (depending on the device and application) include narrow spectral sensitivity, low leakage, significantly differing hole-electron ionization coefficients, and low capacitance.

For some applications, there is a signal-to-noise ratio advantage in obtaining gain within the detecting device itself and avalanching methods are used; in others, there is no advantage to such measures.

REVERSE-BIASED DIODES

Reverse-biased diodes with the depletion region exposed to incident light serve the detection purpose very well, and are employed (with variations) in most receivers, though phototransistors can also be used in applications where speed is not essential. Such devices have relatively broad spectral sensitivity, however, which is a disadvantage for some applications.

• SIMPLE DIODES

Simple pn junction diodes can be usefully employed as detectors (see Figure 3-16). The figure shows an impinging photon liberating a hole-electron pair in the region depleted by the applied reverse bias; each member of the pair individually drifts in response to the field across the depletion region. The propagating charges constitute photoelectric current which is interpreted as received signal.

Those photons which impinge on the undepleted region generate hole-electron pairs which propagate via diffusion, a slower process than drift under appreciable fields. Figure 3-17 illustrates the effects of the diffusion component. The frequency response of simple diodes can thus be seen to be limited.

• P-I-N DIODES

While a conventional p-n diode can serve adequately as a detecting device, the addition of an intrinsic layer provides better control over the depletion depth, improves the collection efficiency, and reduces the diffusion tail.

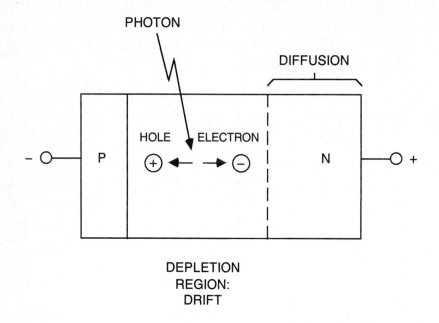

FIG. 3-16. The reverse-biased diode detector.

Structures and Mechanisms

The structure of a p-i-n diode is shown in Figure 3-18, where it will be seen that the depletion region is defined by the size of the intrinsic layer, which depletes under very low bias voltages. Photon collection can be essentially totally limited to the intrinsic region so that diffusion is virtually eliminated as a speed limitation. These diodes are usable at multi-gigabit rates [36].

The quantum efficiency η is defined as the ratio of the captured to incident photons for the device. It improves as the width of the depletion region increases, but the response time improves as this width decreases (on the other hand, a smaller depletion width implies higher capacitance and lower speed). It is therefore necessary to strike a balance between these conflicting requirements in the diode's design and operating conditions.

More specifically,

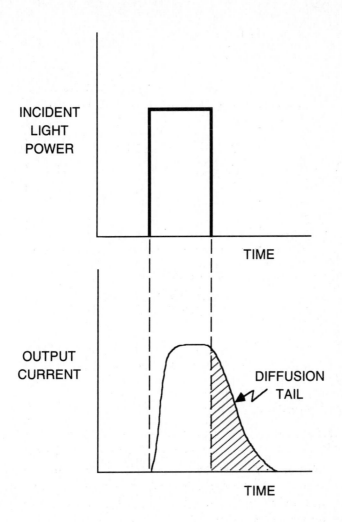

FIG. 3-17. Reverse-biased diode detector response.

$$\eta = \frac{I_p/q}{P_{opt}/h\nu},$$

where

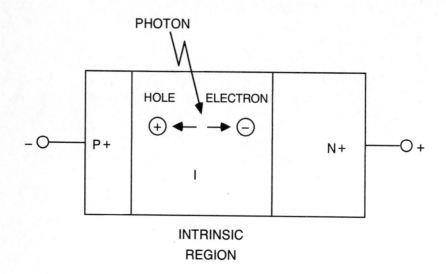

FIG. 3-18. Structure of a p-i-n diode.

I_p is the photogenerated current,

q is the electron charge,

P_{opt} is the incident optical power,

h is Plank's constant, and

v is the frequency of the incident light.

The *responsivity* is a useful figure of merit given by

$$R = \frac{I_p}{P_{opt}} = \frac{\eta q}{h v} \approx \eta \frac{\lambda}{1.24} \qquad \text{Amperes/Watt} \quad (\lambda \text{ in } \mu\text{m}).$$

In modern devices, R is often close to unity, providing, for example, about one microampere of signal current for the typical one microwatt of incident light power available at the extremity of a span.

Capabilities and Limitations

P-i-n diodes can readily be fabricated with quantum efficiencies approaching unity (typically 0.9) and frequency response in the several GHz range is obtainable. The devices are linear over several decades of light intensity [37].

Devices responsive over portions of the spectrum from 0.7 to 1.6 μm are readily available.

AVALANCHE PHOTO-DIODES

The p-i-n devices provide no current gain. For certain applications, additional gain provided at the detector is advantageous.

The Avalanche Photo-Diode (APD) may be thought of as essentially a p-i-n diode that is operated at a reverse bias voltage which is near the avalanche break-down point. Under these conditions, holes and electrons that are produced in the depletion region by an impinging photon are accelerated by the high E-field until their velocity is sufficient to produce secondary hole-electron pairs via collision. Such collisions constitute a random process, introducing additional noise, but the gains can be quite high (of the order of 100 in units of avalanche-produced n-ary hole-electron pairs per captured photon). There is a strong analogy to the operating principle of the photomultiplier tube, except that the collisions in such tubes take place at discrete *dynode* sites rather than on a distributed basis as is true for APDs.

While linear over a wide range of incident optical power, for high light levels (atypical of well-designed systems) APDs will saturate and introduce signal distortion. Their dynamic range is therefore less than that of a p-i-n diode.

Structures

Figure 3-19 indicates the structure of an APD. The high reverse bias voltage induces a wide depletion region comprised of the p, multiplication region, and the intrinsic region. Ideally, photons are captured in the intrinsic region, producing hole-electron pairs, and the electrons drift toward the high-field region. The electrons (which, for example, have a higher ionization coefficient than holes in silicon) are then injected into the multiplication region, where they promote avalanching.

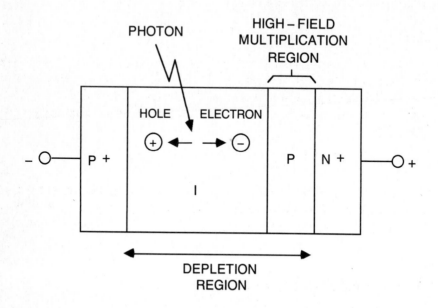

FIG. 3-19. Structure of an APD.

Mechanisms

It is particularly important in APDs to have a ratio of ionization coefficients which is large, i.e., one of the two carrier types should have a significantly greater ability to produce hole-electron pairs via collision. Otherwise, the device will display a large excess noise factor. (Some thought will reveal that, if one of the ionization coefficients were zero, the avalanche process could never be self-sustaining.)

Capabilities and Limitations

APDs can extend the range of fiber transmission systems significantly when the receiver operation is circuit-noise limited. They do, however, require substantially higher bias voltages than p-i-ns (as high as 400 volts), which increases the cost of a receiver, and tends to reduce the life of both photodiode and power supply. Further, diode failure may cause destruction of the immediately following stage of amplification.

MATERIALS

Silicon is utilized in fabricating p-i-ns and APDs for detection of light at wavelengths in the range of 0.7 to about 1.1 μm. For longer wavelengths, which are of increasing interest, germanium and III-V compounds (e.g., InP and GaAs) devices become necessary. Germanium devices also are sensitive at long wavelengths, and have been employed [38], but germanium is not an ideal material because of its high *dark current*, which adds to its noise contribution.

3.3 COUPLING EFFICIENCY

The efficiency with which coupling from a source into a fiber can be accomplished is of interest. A great deal of the generated light power can be lost in inefficient coupling.

BUTT COUPLING

Figure 3-20 depicts the situation when a light source is butted against a fiber end. The light power emitted by a source of area A, with emitted power as a function of θ given by $B(\theta)$ can be expressed as [39]

$$P_{emitted} = 2\pi A \int_0^{\pi/2} B(\theta) \sin \theta \, d\theta \, ,$$

and the light power coupled into the fiber can be obtained by integrating only up to the acceptance angle:

$$P_{coupled} = 2\pi A \int_0^{\sin^{-1}(NA)} B(\theta) \sin \theta \, d\theta \, .$$

The ratio of these two values is the *coupling efficiency*, which, for example, for a Lambertian source with $B(\theta) = \cos \theta$, calculates to $(NA)^2$. A typical NA for a graded-index fiber to which such a source might be coupled would be 0.23, for which the coupling efficiency from the above considerations will be about 5%. It becomes clear that coupling can be a major element in the system loss budget.

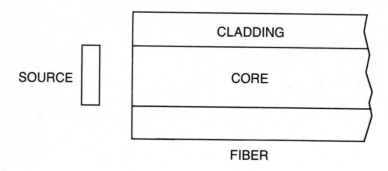

FIG. 3-20. Butt coupling.

LENS COUPLING

The coupling efficiency for sources with poor directionality (e.g., surface-emitting LEDs) can be greatly improved with the aid of an appropriate lens at the added cost of the lens itself, its mount, and any alignment procedure. Spherical lenses are favored because they are relatively inexpensive to manufacture and require only simple alignment. (Review Figure 3-11.)

TAPERED FIBER

Early experiments conducted [40] utilizing fiber that had been tapered down from the standard size to a significantly smaller diameter at the end to be butted (see Figure 3-21) proved successful in improving coupling, and this method has reached commercial practice.

ROUNDED FIBER

Fiber whose end has been rounded to serve as a partial lens is sometimes used as an aid to coupling (see Figure 3-22).

3.4 SAFETY CONSIDERATIONS [41]

The light emitted by the launching devices discussed may, under certain circumstances, present potential risk to the eye. Normally the output at a connector or the end of a fiber is divergent; typical divergence angles are of the order of 10 degrees for multimode fibers, and 6 degrees for single-mode fibers. Thus

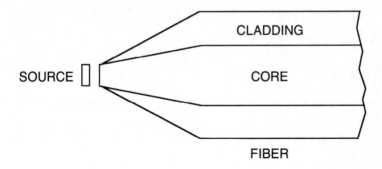

FIG. 3-21. Tapered-fiber coupling.

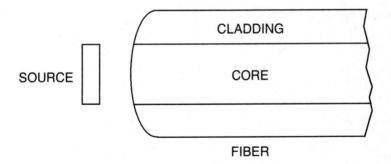

FIG. 3-22. Rounded fiber end.

the irradiance decreases with distance, and at normal viewing distances only a few percent of the total radiated power enters the eye. Under typical viewing conditions, therefore, inadvertent viewing with the unaided eye normally will not cause damage. However, if optics are used to view the emissions of some lens-connectors, there may be some risk. The Maximum Permissible Exposure (MPE) for a given source is a function of the power output, the wavelength, and the duration of the event. Diffuse reflections may also be dangerous for some classes of sources.

At visible wavelengths, there is some intrinsic warning of emission, but at longer wavelengths, the presence of emissions cannot be detected by the human eye.

ILDs are generally more dangerous than LEDs because of their more intense and concentrated flux, but LEDs can also be dangerous if viewed through optical equipment.

There exist standards for handling such devices, e.g., ANSI Z136.1, 1986: the American National Standard for the Safe Use of Lasers, published by the American National Standards Institute.

3.5 RADIATION EFFECTS

Neutron radiation is the primary culprit in producing harmful irradiation effects in light sources and detectors. Because it produces additional recombination centers in the irradiated material, it has the effect of shortening minority carrier lifetime. This reduces the probability of radiative recombination in emitters, and recovery of minority carriers in advantageous regions of detectors.

Radiation is not necessarily deleterious to all devices. Irradiation has actually been observed to increase the light output from one class of LEDs via creation of additional radiative centers [42]. On the whole, however, radiation produces harmful effects in most devices.

More typically, nonradiative recombination centers are created which reduce the output of both LEDs and ILDs. Generally speaking, the effects are less severe in more heavily doped devices, leading to the conclusion that ILDs utilized subthreshold as LEDs provide significantly greater radiation hardness [43] than conventional LEDs.

3.6 SUMMARY

Conversion from the electrical to the optical domain for communications purposes can best be performed using LEDs or lasers (ILDs or traditional types [e.g. ruby or gas]). Lasers derive their performance characteristics from the interaction of photons and electrons in elevated energy states. Electrons may be elevated via pumping as in gas lasers or via injection across a forward biased p-n junction in a semiconductor. A photon of a matching energy level impinging upon an elevated-state electron will stimulate it to revert to its ground state and emit a photon of the same wavelength, in phase with and propagating in the same direction as the impinging photon.

In a classical Fabry-Perot laser, a "cavity" promoting standing-waves via partial reflection at the cavity ends is utilized. Semiconductor injection laser diodes form optical cavity boundaries via cleaving, utilizing heterojunctions, ion bombardment, etc. Variations in the interests of reducing the number of longitudinal modes supported (and therefore the line-width) include DFB structures and coupled-cavities.

ILDs are capable of launching powers higher than 10 dBm with linewidths as small as 0.01 nm (single longitudinal mode), though typical values are of the order of 0 dBm and 10 nm or more, respectively.

LEDs fall into the categories of surface emitters and edge emitters. Surface emitters produce light which emanates from a relatively broad surface, producing a Lambertian radiation pattern. Edge emitters are basically ILD-like structures held below the lasing threshold.

Surface emitters generally respond to lower modulation frequencies but provide higher power output than edge emitters. They are relatively ineffective (though not useless) for launching light into single-mode fibers, but can launch significant power into graded-index fiber, especially when aided with an appropriate (typically spherical) lens.

Edge emitters, though their output is less directional than ILDs, can launch sufficient power into single-mode fiber to be effective for short-span links.

Visible-light LEDs are of little utility in communications systems except as launchers for certain plastic fiber.

The materials employed in light-emitting devices are typically combinations of III-V materials proportioned to tune the output wavelength appropriately while maintaining an adequate crystallographic match.

For optical to electrical conversion, reverse-biased diodes are typically employed. P-i-n diodes provide excellent frequency response and linearity but are lossy. APDs are biased close to breakdown so that avalanching can be used as a gain mechanism. Both types detect light as the result of impinging photons inducing hole-electron pair generation, with the consequent collected charge joining the leakage current as a photoelectric component.

Coupling between light-emitting devices and fiber can be via simple butting or with the aid of lenses. The fiber itself may have its end prepared to function as a lens (or may be tapered) for improved coupling.

The laser types used for communications are relatively harmless to human tissue in normal usage, but have potential for damage.

The semiconductor structures used as light emitters and detectors are vulnerable to the same radiation damage mechanisms as other semiconductor devices.

EXERCISES

1. Consider an ILD for which the interior loss is 3 dB per mm, and which has facets both of which reflect one-half the incident light. (Actual GaAs-Air interface is R $\approx$ 0.32.) If the device were 100 μm long, how much gain per unit length would be required to sustain lasing?

2. Find the critical angle within a typical ILD structure with cavity index of refraction $n_1 = 3.6$ and adjacent layer index of refraction $n_2 = 3.4$.

3. Find the quantum efficiency for a p-i-n detector which produces 0.5 μA of signal current for a 1 μW incident light signal at 1 μm. What would its responsivity be?

4. Diagram a progression of events that would illustrate an infinite decay-time in an APD when the ionization coefficients of holes and electrons are equal.

5. Assuming a potential of 5 V across the intrinsic region of a silicon p-i-n diode with a thickness of 5 μm, what would be the maximum transit time for an electron generated (along with a hole) by an impinging photon of appropriate λ? Assume a mobility of 10^3 cm^2 per volt-second.

6. If nonlinear Raman scattering occurs at power densities above 10^8 watts per square cm, what is the maximum power level that a laser may be permitted to launch into a single-mode fiber with a 10 μm diameter core if such scattering is not permissible? If this scattering were the limiting effect, what would be the maximum unrepeatered range of a fiber system assuming no splices?

7. Find an expression for the coupling efficiency for a source with the generalized radiation pattern $B(\theta) = (cos\ \theta)^n$.

8. Consider the feasibility of designing a single device intended to serve as *either* an LED or a detector diode. If forward biased, it should serve as a respectable light source, and if reverse biased, it should have a usable responsivity. How would you propose fashioning such a device? What compromises would have to be made? What materials would be

required? What uses would such a device have?

9. Outline what precautions you would take with regard to a 10 mW Helium-Neon laser to be used routinely in a laboratory environment. How would you shield casual visitors from exposure? What training would you provide to other users? How would you handle the device yourself?

REFERENCES

[1] H. J. Round, "A Note on Carborundum," *Electron World*, vol. 19, 1907, p. 309.

[2] J. P. Gordan, H. J. Zeiger, and C. H. Townes, "Molecular Microwave Oscillator and New Hyperfine Structure in the Microwave Spectrum of NH3," *Phys. Rev.*, vol. 95, 1954, p. 282.

[3] T. H. Maiman, "Stimulated Optical Radiation in Ruby Masers," *Nature*, vol. 187, 1958, p 493.

[4] R. N. Hall, G. E. Genner, J. D. Kingsley, T. J. Soltys, and R. O. Carlson, "Coherent Light Emission from GaAs Junctions," *Physics Review Letters*, vol. 9, 1962, p. 366.

[5] N. Holonyak, Jr. and S. F. Bevelacqua, "Coherent (Visible) Light Emission from Ga(As[1-x]P[x]) Junction," *Applied Physics Letters*, vol. 1, 1962, p. 82.

[6] H. Kroemer, "A Proposed Class of Heterojunction Injection Lasers," *Proceedings IEEE*, vol. 51, 1963, p. 1782.

[7] I. Hayashi, M. B. Panish, P. W. Foy and S. Sumelay, "Junction Lasers Which Operate Continuously at Room Temperature," *Applied Physics Letters,* vol. 17, no. 3, 1970, pp. 109-111.

[8] W. E. Kock, *Engineering Applications of Lasers and Holography,* p. 331.

[9] J. M. Carroll, *The Story of the LASER,* E. P. Dutton & Co., 1964, p. 123.

[10] G. W. Flint, "Analysis and Optimization of Laser Ranging Techniques," *IEEE Military Electronics,* January, 1964, p. 22.

[11]   Ettenberg and H. Kressel, *Journal of Applied Physics*, vol. 47, 1976, p. 1538.

[12]   H. Kressel, *Topics in Applied Physics: Semiconductor Devices for Optical Communication*, Springer-Verlag, 1980, pp. 54-55.

[13]   H. Kroemer, "Heterostructure Bipolar Transistors and Integrated Circuits," *Proceedings of the IEEE, Special Issue on Very Fast Solid-State Technology*, vol. 70, no. 1, January, 1982, pp. 13-25.

[14]   H. Kogelnik and C. Shank, "Coupled-Wave Theory of Distributed Feedback Lasers," *Journal of Applied Physics*, vol. 43, 1972, p. 2327.

[15]   C. S. Tsai, "Guided Wave Acoustooptic Bragg Modulators for Wide-Band Integrated Optic Communication and Signal Processing," *IEEE Transactions on Circuits and Systems*, vol. CAS-26, no. 12, December, 1979, pp. 1072-98.

[16]   G. P. Agrawal and N. K. Dutta, *Long-Wavelength Semiconductor Lasers*, Van Nostrand Reinhold, 1986, p. 289.

[17]   W. T. Tsang and R. Logan, *IEEE Journal on Quantum Electronics*, vol. QE-19, November, 1983.

[18]   W. T. Tsang and N. A. Olson, "High-Speed Direct Single-Mode Modulation," *Applied Physics Letters*, vol. 42, 1983, p. 650.

[19]   "New Laser Seen as Major Advance in Lightwave Communications," *AT&T Bell Laboratories Record*, April, 1983, p. 2.

[20]   R. Linke, B. Jasper, J. Ko, I. Kaminow and R. Vodhanel, *Proceedings of the 4th International Conference on Integrated Optics and Optical-Fiber Communications*, Tokyo, 1983.

[21]   R. E. Epworth, "The Phenomenon of Modal Noise In Analogue and Digital Optical Fiber Systems," *4th European Conference on Optical Communications*, Genoa, 1978.

[22]   C. A. Burrus and R. W. Dawson, "Small-Area High-Current Density GaAs Electroluminescent Diodes and a Method of Operation for Improved Degradation Characteristics," *Applied Physics Letters*, vol. 17, 1970, pp. 97-98.

[23] I. P. Kaminow, G. Eisenstein, L. Stulz and A. Denki, "Lateral Confinement in InGaAsP Super-Luminescent Diode at 1.3 Microns," *IEEE Journal on Quantum Electronics*, vol. QE19, no. 1, January, 1983, pp. 78-82.

[24] G. Arnold and O. Krumpholz, "Coupling of Monomode Fibers to Edge-Emitting Diodes," *Optical Fiber Communications/Optical Fiber Sensors*, February, 1983, pp. 48-49.

[25] G. Arnold et al., "1.3 Micron Edge-Emitting Diodes Launching 250 Microwatts into a Single-Mode Fibre at 100 mA," *Electronics Letters*, vol. 21, 1985, pp. 993-994.

[26] Plastow et al., "Transmission Over 107 km of Dispersion-Shifted Fibre at 16 Mbit/s Using a 1.55 Micron Edge-Emitting Source," *Electronics Letters*, vol. 21, 1985, pp. 369-370.

[27] *Electronics Week*, (PlessCor Optronics, Inc.), January 21, 1985, p. 25.

[28] S. M. Sze, *Physics of Semiconductor Devices, 2nd Edition*, John Wiley and Sons, 1981, pp. 689-700.

[29] J. Gosch, "Silicon Carbide Ends Long Quest for Blue Light-Emitting Diode," *Electronics Week,* October 8, 1984, p. 24.

[30] Kressel, p 10.

[31] H. C. Freyhardt, Ed., *Crystals: Growth, Properties, and Applications*, Springer-Verlag, 1980, vol. 3, pp. 73-162.

[32] R. D. Dupuis, L. A. Moudy, and P. D. Dapkus, "Preparation and Properties of Ga(1-x)Al(x)As-GaAs Heterojunctions Grown by Metal-Organic Chemical Vapor Deposition," *Gallium Arsenide and Related Compounds 1978*, Inst. Phys. Conf. Ser., vol. 45, pp. 1-9.

[33] *Electronic Engineering Times*, May 20, 1985, pp. 27ff.

[34] T. P. Lee, C. A. Burrus, and A. G. Dentai, "Dual Wavelength Surface Emitting InGaAs L.E.D.s," *Electronic Letters*, October 23, 1980, vol. 16, no. 22, pp. 845-846.

[35] K. Ogawa, T. P. Lee, C. A. Burrus, J. C. Campbell, and A. G. Dentai, "Wavelength Division Multiplexing Experiment Employing Dual-Wavelength LEDs and Photodetectors," *Electronics Letters*, October 29,

1981, vol. 17, no. 22, pp. 857-858.

[36]   L. A. Godfrey and B. K. Garside, "Optimal Design of Ultrafast Photo-detectors," *Conference on Quantum Electron Lasers and Electronics,* June, 1981, p. 66.

[37]   Kressel, p. 66.

[38]   Ando et al., "Characteristics of Germanium Avalanche Photodiodes in the Wavelength Region of 1-1.6 Microns," *IEEE Journal of Quantum Electronics,* November, 1978, pp. 804-9.

[39]   M. K. Barnowski, *Fundamentals of Optical Fiber Communication,* Academic Press, Inc., 1972, p. 92.

[40]   Y. Yamatsu and T. Ozeki, *Technical Digest of 1977 International Conference on Integrated Optics and Optical Fiber Communications,* p. 371.

[41]   R. C. Peterson and D. H. Sliney, "Toward the Development of Laser Safety Standards for Fiber-Optic Communication Systems," *Applied Optics,* vol. 25, no. 7, April 1, 1986, pp. 1038-1047.

[42]   C. E. Barnes, *Radiation Effects in Optoelectronic Devices,* Sandia Report, SAND76-0726, 1976, p. 6.

[43]   Ibid, p. 97.

CHAPTER 4

CIRCUITS: TRANSMITTING AND RECEIVING

4.1 CIRCUITS

The circuits necessary to effectively drive light-emitting devices and to amplify the feeble signal presented to the photodetecting devices are somewhat unique to the fiber optics world. Because of the inefficiencies of transduction and other losses, a remarkable difference exists between the electrical power delivered to the emitting device and the electrical power available from the detecting device. These differences are reflected in the makeup of and problems associated with the amplifiers at either end of a fiber link.

4.2 TRANSMITTING CIRCUITS

The nature of the transmitter driving circuits is strongly affected by whether the driven device is an LED or an ILD, because these devices behave in significantly different ways. One attribute they share is the need to muster drive currents typically of the order of 100 ma or more (peak currents of much higher value may be used). Only a relatively small subclass of circuits is capable of providing current levels in this range at the speeds called for in many fiber optics systems.

LED DRIVE CIRCUITS

Surface-emitting LEDs are relatively easy to drive when low frequency operation is adequate. The major requirement of such circuitry is high-current capability. Temperature stabilization of the LED itself is usually unnecessary, but the circuit dissipation may be great enough to require heat-sinking capability, especially if such diodes are fabricated in monolithic arrays. Figure 4-1 depicts a typical circuit. This circuit has the advantage of precise current control

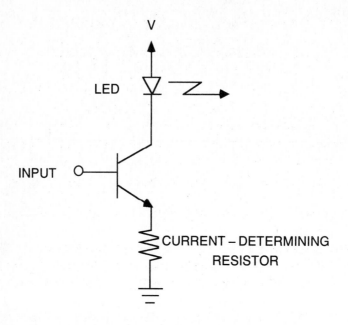

FIG. 4-1. A typical LED drive circuit.

independent of the diode drop.

From a driving circuit design point of view, edge-emitting LEDs may be regarded as compromises between surface emitting LEDs and ILDs, sharing some of the temperature stabilization problems of ILDs.

Some attention must be paid to the "off" bias point if high-speed operation is desired. If the LED is reverse biased, the junction capacitance must be discharged before the junction can be forward biased again, requiring current and time. It is therefore necessary to bias the LED slightly "on" in the nominally "off" state for high-speed response. Unfortunately, this practice impairs the *extinction ratio* (the ratio between light intensity for a *mark* and a *space*), reducing the signal-to-noise ratio.

Analog Transmission

Analog applications tend to be less numerous than digital ones, except for relatively low frequencies and single-channel operation. Though significantly more linear than typical ILDs, LEDs display sufficiently nonlinear behavior to produce disturbing intermodulation distortion products (due to mixing) among multiple channels. Transmission of a single NTSC (National Television Standards Committee) video channel is quite feasible, for example, but frequency multiplexing several channels creates problems unless carrier frequencies are chosen with great care to reduce the problem of intermodulation distortion [1]. Channel capacities rivaling those of coaxial cable systems appear to be unattainable, at least for transmission at a single wavelength using incoherent techniques. (Wavelength division multiplexing offers significantly greater capacity, as do coherent techniques.)

Dissipation can present a more significant problem for analog circuits because the drive circuit functions in the active region (neither saturated nor cutoff) continuously, and the time integral of the I-V product may be significant.

In general, circuitry which provides current independent of the diode's voltage drop is favored for analog applications in order to prevent compounding the innate nonlinearity of light-launching devices.

Digital Transmission

Digital transmission places no constraints upon linearity, so that transmitting circuits can be relatively simple and of low dissipation because they function as digital gates, with transistors spending little time traversing the highly dissipative active region, mostly residing in high-current, low-voltage saturation conditions, or high-voltage, virtually zero-current, cut-off conditions. As bit rates are increased, however, the proportion of time spent in the highly dissipative active region increases, in turn increasing the average dissipation of the driving device. Figure 4-2 displays a typical digital drive circuit for an LED. The form of a TTL logic gate will be recognized, comprising the single-input remnant of the logic-performing front end followed by a *phase splitter* driving a *totem-pole* output. Such gates are usually designed to have substantial current sinking capability but relatively indifferent current sourcing capability because, when driving gates of the same kind, after charging parasitic capacitance on pull-up, no current is drawn, while significant DC current is drawn after pull-down. Thus, the gate can be effective in a configuration that sinks current from an LED rather than sourcing current to it.

The speed performance of digital systems in transmission applications is often expressed in Megabits per second (Mbps) with a Non-Return to Zero (NRZ) format assumed. It can be shown [2] that the bandwidth necessary to support a given NRZ bit rate is numerically one-half that value; that is, up to

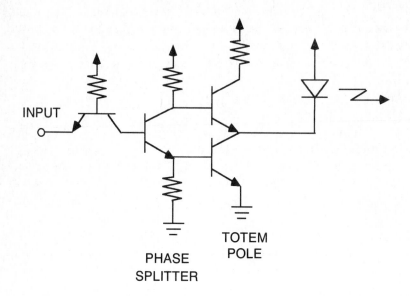

FIG. 4-2. A typical digital LED drive circuit.

two bits per second can be conveyed per Hz of bandwidth.

Figure 4-3 indicates an alternative circuit which supplies "on" current via a resistor when the transistor is cutoff and shunts the resistor current away from the diode for the "off" condition via saturating the transistor.

ILD DRIVE CIRCUITS

ILD drive circuits are considerably complicated by the need for feedback means capable of adjusting the value of the bias current as the threshold changes with temperature and age. Further, the ILD devices tend innately to be highly nonlinear, making them suitable primarily for digital or certain varieties of pulse-analog operation. Relatively linear regions of some ILDs may, however, be usefully employed for analog purposes.

Since ILDs are capable of modulation at several GHz, quite exotic combinations of drive circuits and devices may be necessary.

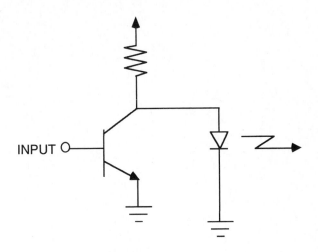

FIG. 4-3. An alternative implementation.

ILD Stabilization

Stabilization of the operating point of an ILD under temperature fluctuations and aging can be performed in several different ways, classifiable upon whether they monitor light output or temperature or are purely electronic, which facet is monitored, etc.

• *Light Monitoring*

Double-faceted Fabry-Perot structures provide two choices for monitoring the ILD light output. Most commonly, light emanating from the rear facet is monitored with an in-package p-i-n diode (see Figure 4-4). While this approach is satisfactory for most purposes, under some conditions the back-face light output is not a sufficiently reliable mirror (no pun intended) of the behavior of the front-face output. The other choice is to monitor the front-face output via a tap (see Figure 4-5).

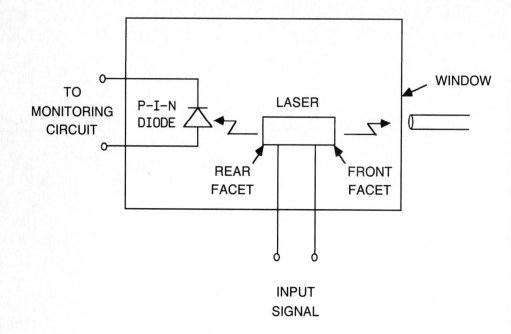

FIG. 4-4. Rear facet monitoring.

• *Temperature Control*

In essentially the same fashion as quartz crystals may be kept at
constant temperature to stabilize oscillation frequency, the temperature of an ILD may be kept stable. Thermo-electric coolers are
commonly used in such applications, and can readily keep the temperature constant to within $1C^o$. Substantial current is drawn by
such devices, however, contributing to an overall low efficiency.

• *Electronic Sensing*

Figure 4-6 shows a typical laser-diode I-V characteristic. A careful
scrutiny of the figure will reveal that there is a point of inflection at
threshold which holds the potential of yielding to electronic sensing.
This can be illustrated from the change in behavior of the device at
this point: below threshold, the device I-V characteristic functions

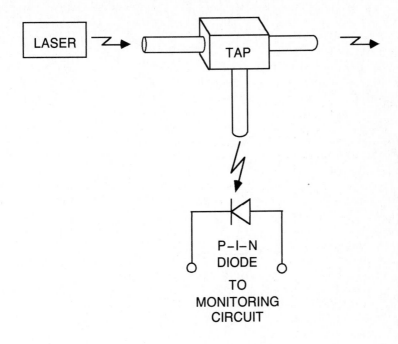

FIG. 4-5. Front facet monitoring.

as that of a diode:

$$I = k_1 \left[e^{k_2(V - IR)} - 1 \right],$$

where R is the device resistance, and the k's are proportionality constants.

Above threshold, the junction voltage is *pinned* at the bandgap value (plus the IR drop):

$$V = E_g/e + IR.$$

To a first approximation, the I-V characteristic can be seen as changing from an exponential to a linear behavior at threshold.

Two possible techniques for taking advantage of this property for threshold sensing are shown in Figures 4-7 and 4-8. In the first

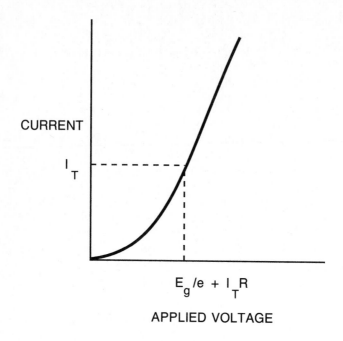

FIG. 4-6. ILD I-V characteristics.

figure is a circuit which senses the pinning with a bridge circuit on a DC basis [3]; in the second figure is a circuit which senses threshold by monitoring the presence or absence of harmonics generated by the response to a pilot signal applied continuously out of band, or periodically in-band. The advantages of purely electronic sensing are that no additional, positioning-critical optical component is required, no expensive, high-current-draw thermo-electric cooler need be employed, and though the circuitry may become complex, it is amenable to VLSI.

Analog Transmission

As stated previously, ILDs are rarely employed for analog transmission because of their severely nonlinear behavior. Some ILDs have been produced which are substantially more linear (above threshold) than most. On the whole, however, they are best relegated to digital transmission.

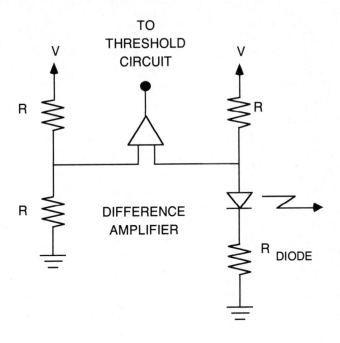

FIG. 4-7. DC sensing of pinning.

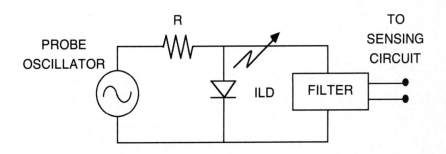

FIG. 4-8. AC sensing of pinning.

Digital Transmission

ILDs tend to be employed in very high-rate digital applications requiring nimble, high-powered drive circuitry. The high speed increases the peak current demands because parasitic capacitance must be rapidly charged and discharged. Parasitic inductance also increasingly becomes a problem at elevated frequencies.

Figure 4-9 shows a circuit appropriate to rates of the order of 400 Mbps. It is basically an Emitter-Coupled Logic (ECL) circuit, an example of what is often called Current-Mode Logic (CML). The philosophy of this circuit type is to function as a switched, constant-current source or sink in order to charge capacitance rapidly. The transistors are kept out of saturation in the interest of speed. The current level is determined by the emitter resistor value and reference voltage (V_{REF_1}). The constant current draw results in relatively "quiet" power supply busses, but also assures high dissipation.

For higher frequencies ($\geqslant$ 1 Gbps), the same circuit may be employed with more specialized discrete devices (e.g., interdigitated emitter structures such as those used in microwave transistors) and, most likely, with a balanced drive, obviating the need for the second reference potential.

At still higher frequencies, higher mobility semiconductor materials systems (e.g., GaAs [4]) for transistor fabrication become important, as may heterojunction structures [5].

LINEARIZATION

Several techniques can be used for overcoming the nonlinear properties of light-launching devices.

• Feedback Techniques

As is true for purely electrical circuits, feedback techniques can be used to linearize the light-power versus current characteristics of LEDs and ILDs. Since it is not practical to sense the instantaneous light power output via exclusively electronic means, the feedback loop must include a photonic device to sample the light and convert the sampled value to a corresponding electrical level.

Figure 4-10 shows a representative circuit capable of reducing the distortion of a light-launching device. As is true of any feedback circuit, the frequency response is limited by the delay around the loop, in turn limiting the circuit to relatively nondemanding

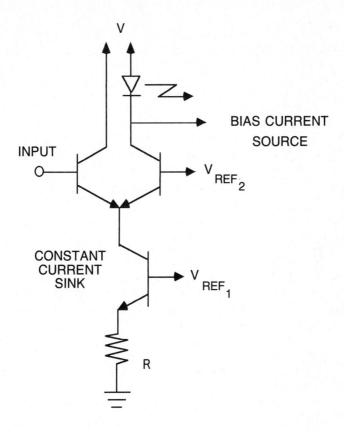

FIG. 4-9. A high-speed drive circuit.

applications (typically, the amplifier must have some ten times the bandwidth of the input signal).

● *Feedforward Techniques*

Feedforward techniques, long overshadowed by feedback techniques [6, 7], can provide considerably higher frequency response than feedback approaches. Figure 4-11 displays the classical form of the circuit.

The uppermost of the two amplifiers in the figure amplifies the input signal, contributing distortion and noise components. The

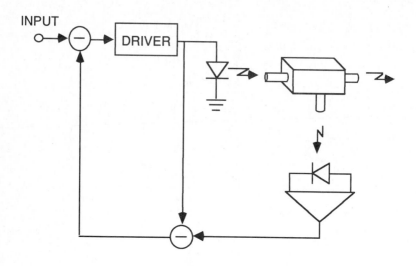

FIG. 4-10. Feedback linearization.

lower summing circuit subtracts the input signal from an appropriately attenuated version of the corrupted signal, extracting the distortion and noise. This correction signal is elevated in level by the
second amplifier, and the ultimate summer subtracts this component
from the corrupted signal, ideally yielding an unsullied, amplified
image of the input signal.

It will, of course, be recognized that the second amplifier will
contribute its own burden of distortion and noise, but these are
third-order perturbations on the second-order distortion and noise
signals.

The implementation of a feedforward photonic system is left as
an exercise.

● *Frequency Choice*

The intermodulation products produced by nonlinearities are deter-
ministic and calculable as a function of the signal spectra. Spurious
distortion components can be rendered harmless by avoiding their
portion of the spectrum. While wasteful of bandwidth in a sense,
there is a great deal of that commodity to spare in many fiber

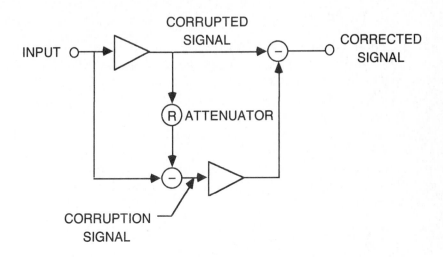

FIG. 4-11. Feedforward linearization.

systems.

Such techniques have proven effective in some applications, leading to transmission, for example, of 12 video channels with the same light-launching device.

Another approach, using a phase-correlated frequency scheme, was successful in transmitting 26 video channels [8] with the same device.

Prognosis for Linearization Techniques

Though considerable improvement in distortion levels can be attained through circuit techniques, there are limits to the degree of improvement possible. Devices from the same production run may, for example, have differing characteristics, and they may change somewhat with time. It is possible, however, that self-calibrating linearization circuits capable of correcting arbitrary nonlinearities may become available.

4.3 RECEIVER CIRCUITS

Receivers represent perhaps the most complex and demanding part of the design of a lightwave transmission system. Received signals, though virtually uncorrupted in transit over the fiber span, are typically of the order of a microwatt or less in amplitude, so that all of the considerations associated with detecting feeble signals in the presence of noise must be brought to bear.

The noise sources in the receiver are several: thermal or Johnson noise (associated with resistive elements), shot noise (resulting from the discrete nature of the arriving photon energy), dark current noise (due to the leakage current in the detecting device), and device noise (associated with the active devices employed).

Some of the noise sources are signal dependent and some are not. Depending upon the constraints of specific applications, there is some latitude for choice among alternative circuit structures.

To those accustomed to digital circuits, it may come as a shock when it is recognized that the amplitude of a received optical signal is far too low for any reasonable circuit to interpret digitally. The early stages of the receiver are therefore necessarily analog, and essentially linear.

The detector diodes, whether p-i-n or APD, are reverse biased, and present an impedance to the preamplifier which is virtually purely capacitive [9].

Figure 4-12 is a block diagram of a typical receiver. The photodetector is followed by a preamplifier whose task is to bring the signal power up to a moderately high level with minimum added noise. Two major categories of preamplifier will be discussed.

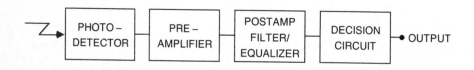

FIG. 4-12. A typical receiver structure.

Postamplification, filtering to narrow the noise bandwidth, and equalization circuits may follow the preamplifier. If the transmission is digital, a discrimination circuit is typically employed. Timing recovery circuitry may or may not be employed, and while in the domain of the application, is beyond the scope of this book (some of its aspects will, however, be discussed in association with scrambling techniques and in one of the exercises).

PREAMPLIFIERS

The preamplifier must be designed to meet the application requirements with respect to *noise figure* and *dynamic range*. These parameters are particularly critical when the conveyed signal is analog.

Fortunately, the nature of the receiving device aids in signal-to-noise ratio considerations relative to the incident signal: because the receiving devices produce an electrical *current* which is proportional to the incident light *power*, the corresponding electrical power is proportional to the *square* of the incoming light power. The electrical-domain signal-to-noise ratio in dB is therefore double the light-domain signal-to-noise ratio in dB. Thus, for example, a 15 dB signal-to-noise ratio in the light domain would correspond to a 30 dB signal-to-noise ratio in the electrical domain if the detecting process contributed no additional noise.

The Transimpedance Preamplifier

The basic transimpedance amplifier is shown in Figure 4-13. It may be viewed as a "poor man's" operational amplifier.

In a conventional operational amplifier, the open-loop amplifier gain is very large in order to obtain characteristics approaching those of an ideal circuit. At the frequencies typically being transmitted over fiber facilities, however, the delay around the feed-back loop of a conventional operational amplifier becomes so great that operational behavior is not possible (with some exceptions, conventional operational amplifiers become uncomfortable much above 1 MHz). It is therefore necessary to construct a circuit which provides high speed around a temporally short feedback loop at the expense of some gain; hence the transimpedance amplifier.

The philosophy of the transimpedance amplifier is implicit in its name. Power amplification is obtained in response to an input current signal (the photoelectric current in the detector resulting from collection of the hole-electron pairs produced by annihilated photons) by producing a corresponding voltage output at much lower impedance, yielding the potential for a large current signal. The transfer characteristic of the amplifier is the ratio of the output voltage to an input current and is therefore expressible as an impedance.

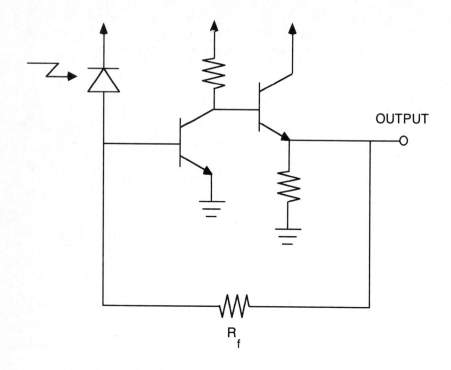

FIG. 4-13. The transimpedance amplifier.

Examining Figure 4-13, it will be seen to comprise a first stage which is a high-gain common-emitter amplifier, and a second stage which is an emitter follower (in bipolar parlance; common source and source follower, respectively, for unipolar devices). The first stage ideally provides high voltage gain and inversion, the second stage, high current gain without phase-shift. A feedback resistor completes the negative feedback loop.

To a first order, the behavior of such a circuit can be analyzed as follows (see Figure 4-14):

> For the gain of the first stage equal to -A (explicitly indicating the inversion),

$$V_3 = V_2 = -AV_1 .$$

Assuming that the input impedance of the first stage is large

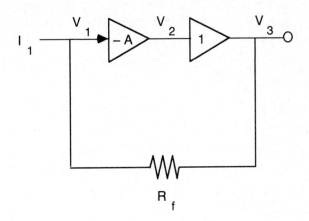

FIG. 4-14. Transimpedance amplifier analysis.

compared to $\dfrac{R_f}{1 + A}$, all of I_1 must traverse R_f, so that

$$I_1 R_f = V_1 - V_3,$$

$$= -\frac{V_3}{A} - V_3 = -V_3 \left[\frac{1 + A}{A} \right].$$

Therefore,

$$-\frac{V_3}{I_1} = \frac{A R_f}{1 + A},$$

and, for A very large,

$$\frac{V_3}{I_1} = -R_f.$$

In real circuits, of course, the assumptions made above are only approximations:

- The amplifier was assumed to be strictly unilateral (transistors, in general, are not perfectly unilateral, so that non-obvious parasitic feedback loops also exist [10]),

- The input impedance of the first stage is finite, given (approximately) by βr_e, and β is a function of frequency,

- The gain of the first stage is by no means infinite, and drops at elevated frequencies,

- The phase shift of the first stage does not remain the idealized 180° nor that of the second stage at 0° at very high frequencies (net phase may vary more than 100° over the band of interest [11]), and

- The delay around the feedback loop, which represents phase shift, becomes a greater proportion of a cycle at higher frequencies.

Thus, the transimpedance amplifier is nonideal, but it has other characteristics which make it attractive:

- It has a wide *dynamic range* (that is, it can amplify signals over a range of several decades in size without saturating). A well-designed system using a transimpedance amplifier may function essentially as well when the fiber is removed and the transmitter and receiver are butted together.

- It can be readily integrated onto an IC chip (including the light detector if the wavelength to be used permits the detector to be of the same semiconductor system as the transistors; this is particularly true at short wavelengths, where silicon diodes may be used, but at long wavelengths, either germanium or III-V compound detectors must be employed, neither of which is yet ideal for integration with the preamplifier, though progress is being made).

- It does not fundamentally require equalization circuits to follow it, though for extreme applications some equalization may be called for.

The noise introduced by the feedback resistor in this circuit is thermal, with a mean-square value of $<i^2> = \dfrac{4kT}{R_f}\Delta f$, where T is temperature, k is Boltzmann's constant, and Δf is bandwidth. In some modern implementations, FET preamplifiers are used (coining the term PINFET when p-i-n diodes are used [see Figure 4-15]). In such circuits, the role of the feedback resistor is often also played by an FET, reducing the noise figure because of the low-noise properties of FETs compared to equivalent-resistance resistors. The drain load of the first stage and the source load of the second stage are also typically implemented with depletion-mode FETs as shown in the figure.

The value of transimpedance gain for practical circuits may be anywhere from 1,000 ohms to several Megohms.

An interesting extension of the basic transimpedance preamplifier has been successfully tested at low bit rates. It employs an optical feedback mechanism to replace the conventional feedback path, eliminating the thermal noise and feedback capacitance associated with a feedback resistor, and providing a wide dynamic range [12].

Integrating Amplifier

The amplifier of Figure 4-16 is referred to as an *Integrating Amplifier*. Instead of presenting a low-impedance point to the signal current of the detecting device as did the transimpedance amplifier, the high-impedance of an unencumbered transistor base (or gate) is presented to the input signal. The combination of a high impedance and the capacitance of the reverse-biased detector diode and transistor base constitute a low-pass (integrating) structure that greatly limits the response of the front-end. Fortunately, however, the response can be compensated later via an equalization circuit with complementary highpass characteristics after the signal has been brought to reasonable impedance levels (see Figure 4-17 [13] for an illustration of the remarkable effectiveness of post-amplification equalization).

It can be shown that the integrating amplifier has signal-to-noise characteristics somewhat superior to those of the transimpedance amplifier [14], but its dynamic range is significantly less, so that its gain must be "tuned" to the application. The equalization circuit, similarly, should have its elements chosen as a function of the detector parasitics, and may therefore, in practice, require adjustability.

For most practical applications, the transimpedance amplifier is preferred to the integrating amplifier.

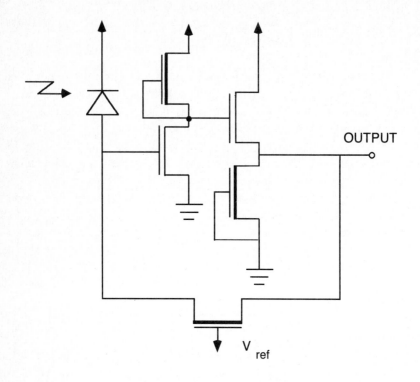

FIG. 4-15. The pinfet preamplifier.

Signal-to-Noise Ratio

It will be recalled that signal-to-noise ratio (S/N) is a power ratio usually expressed in dB. The parameter is of interest in this context relative to amplifiers which must deal with very low-level signals that must compete with the noise introduced by the detector-amplifier combination.

The constituents are Johnson or thermal noise of the form $<i^2> = \dfrac{4kT}{R}\Delta f$, which is signal independent; shot noise of the form $<i^2> = 2qI\Delta f$ (where I is signal current), which is signal dependent; dark-current noise, which is characteristic of the detector, and depends upon its volume and surface area, its constituent material, edge effects and temperature, and is given by $<i^2> = 2qI_D\Delta f$; equalization noise (if it is performed) given by $<i^2> = 4\dfrac{\pi^2}{3}C^2(\Delta f)^3$; so-called *flicker noise* which varies as 1/f, and

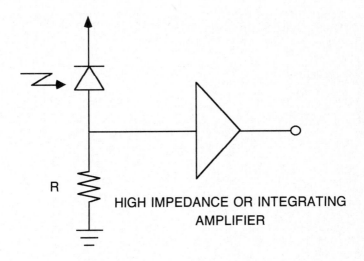

FIG. 4-16. The integrating amplifier.

which may be neglected for most fiber-optics operating frequencies; and noise associated with the amplifier's active devices, which is dependent upon its makeup. (It will be recalled that, since the means of these noise voltages or currents are zero, it is convenient to express them in mean-square form, especially since such values are additive if, as is generally assumed, the noise sources are Gaussian and statistically independent.)

In order to limit the power contributions of the bandwidth-dependent noise sources, band-limiting filters are often employed in receivers to limit the spectral window to frequencies contained in the signal to be amplified. It will be recognized that such amplifiers may be capable of very high frequency operation, but are not *broadband* amplifiers.

The S/N ratio of an amplified signal may be expressed as

$$S/N = \frac{m^2 I^2}{2(<i^2>_{circuit} + 2qI\Delta f)} ,$$

for a p-i-n, ignoring dark current, where m is the modulation index and I is the signal current. For an APD, I is replaced by $I \times M_{mean}$, where M_{mean} is the mean value of the APD gain.

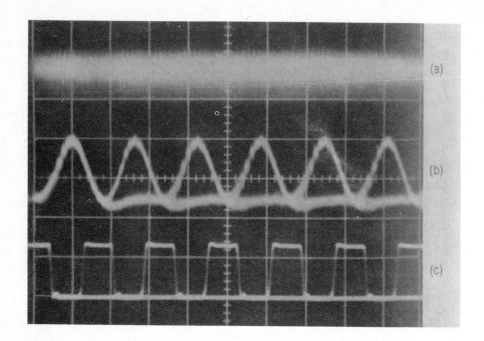

FIG. 4-17. Results of equalization. (Reproduced with
permission, BSTJ, vol. 53, no. 4, April, 1974, p. 640.)

Note that for very small signal current, the S/N ratio is almost solely
dependent upon the circuit noise; this is typical of analog or very long-range
digital transmission.

For large signal current, typical of short-range transmission, the S/N ratio
becomes dependent primarily upon shot noise, and the circuit noise is unimportant.

LIMITS

It is useful to determine ultimate limits upon detectability, given an acceptable
error rate. Consider the random arrival of photons at the detector with an average arrival rate of λ, and assume that the probability of arrival in a given time
subinterval t/k is proportional to that interval:

$$\text{Prob. (arrival in } t/k) = \lambda t/k.$$

Then,

$$\text{Prob. (no arrival in } t/k) = 1 - \lambda t/k \ ,$$

and

$$\text{Prob. (n arrivals in } t) = C(k,n) \ (\lambda t/k)^n \ (1 - \lambda t/k)^{k-n} \ , \qquad (4.1)$$

which will be recognized as the Binomial Distribution.

If the time subinterval is taken to be arbitrarily small:

$$\lim_{k \to \infty} [\text{Prob. (n arrivals in } t)] = \frac{e^{-\lambda t}}{n!}(\lambda t)^n \ , \qquad (4.2)$$

the Poisson Distribution is obtained.

When transmitting Return to Zero (RZ) bits of information with photons, under ideal conditions the light source is completely extinguished when a "zero" is to be transmitted, and the receiver contributes no interfering noise, so that the only error source originates from the lack of at least one photon arriving in the interval allocated to represent a "one". Assuming a Poisson arrival of photons intended to represent a "one", the probability of no arrival in an interval t is given by substituting n = 0 into Eqn. 4.2:

$$\text{Prob. (0 arrivals in t)} = e^{-\lambda t} = e^{-N} \ . \qquad (4.3)$$

For a (typical) specification of a BER of 10^{-9}, this corresponds to N = 21 photons as the quantum limit for representing a logical "one." Assuming ones and zeros are equally likely, this corresponds to 10.5 photons per bit on the average. A more refined analysis concludes that 10 photons per bit suffice for the equally likely case [15].

Clearly, the assumptions made in the above analysis are too extreme for practical direct-detection systems (coherent detection systems, discussed in Chapter 7, may approach this number), but the result may be viewed as a lower bound. Typical values of N needed to represent a "one" in a high-performance system with an adequate error rate is of the order of 1,000 photons.

Analyses utilizing different arrival statistics generally yield similar results [16]; however, there is some hope of altering the actual statistics toward a more favorable arrival distribution [17].

Fig. 4.18 shows a plot of receiver sensitivity in terms of the average signal photons per bit versus bit rate for p-i-n and APD detectors to obtain a 10^{-9} BER. Displayed is the calculated performance of a GaAs FET preamplifier with the two detector types for a total capacitance C_T of one and two picofarads, and no leakage current. The diagonal lines correspond to sensitivity expressed in dBm for a λ of 1.5 μm. Note that the bottom line is at the quantum limit of an average 10 photons per bit.

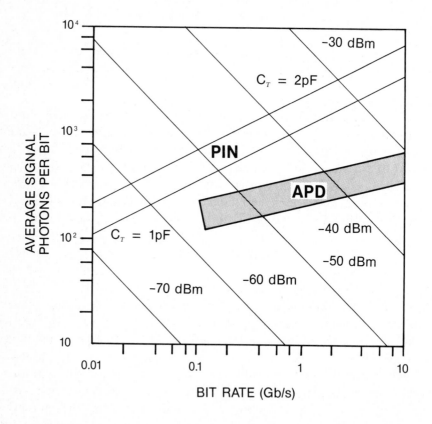

FIG. 4-18. Theoretical receiver sensitivities. (Adapted from [18].)

POSTAMPLIFICATION AND DISCRIMINATION

A postamplifier deals with a signal whose ultimate signal-to-noise ratio has been determined by the devices preceding it. The amplitude and impedance levels that it sees are such that the only significant considerations are the gain it is to provide, the impedance level it must drive, and the bandwidth it is to accommodate. The design of these amplifiers is nonetheless often nontrivial (especially if the bandwidth must extend from almost DC to substantial frequencies as, e.g., is necessary to handle baseband video signals). The topic of postamplifiers stretches the scope of this book and will therefore only be touched upon.

Digital Signals

It will be recalled that a typical received optical power is about one microwatt, and responsivities of detecting devices are often close to unity. The input signal current to a preamplifier may therefore be in the neighborhood of one microampere. The gain of a transimpedance amplifier was found earlier in this Section to be given by

$$\frac{V}{I} = R_f \, ,$$

and R_f is typically chosen to be of the order of 10 kilohms (R_f is chosen to be as large as possible to maximize the S/N ratio, but small enough to provide adequate bandwidth). Thus, the output signal voltage from a transimpedance amplifier may be about 10 millivolts, significantly less than is necessary to trigger standard digital logic gates. A postamplifier is therefore clearly necessary to raise the signal to a suitable level.

A suitably amplified signal is not, however, in general, directly usable as a digital output. The signal may, even after equalization, remain too distorted for presentation to a logic gate with an arbitrary switching threshold. More typically, a discriminator circuit is employed whose threshold is precisely determined (see, e.g., Figure 4-19). The properties of the circuit shown are such that a difference of $\pm$ 0.1 volt at the input relative to the slicing level will switch virtually all of the current from one leg to the other. The constant current draw is advantageous in keeping the voltage supplies clean in proximity to the high-sensitivity preamplifier. The threshold level may be set absolutely or be adaptively adjusted.

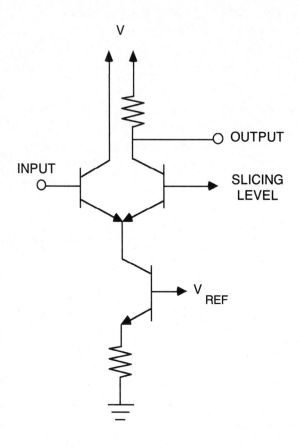

FIG. 4-19. A discriminator circuit.

It is often necessary as well to employ a retiming circuit capable of guaran-teeing the temporal placement of the pulse edges. Such circuits apply phase-locked oscillators to extract the fundamental frequency from a pulse train. For such oscillators to function effectively for arbitrary trains (such as those with extraordinarily long sequences of all zeroes), the signal must be encoded in a form that assures transitions at regular intervals (e.g., *scrambled*). Manchester encoding (see Figure 4-20) is an example of a technique that guarantees regu-lar transitions. (Such techniques are also necessary when AC-coupled amplifiers are used.)

As will be seen, the Manchester algorithm compels the signal to make the transitions in the direction of the represented bit value at each bit position, with additional transitions as required to complete a continuous signal. An Exclusive-Or logic function with a clock is adequate for conversion to or from this encoding.

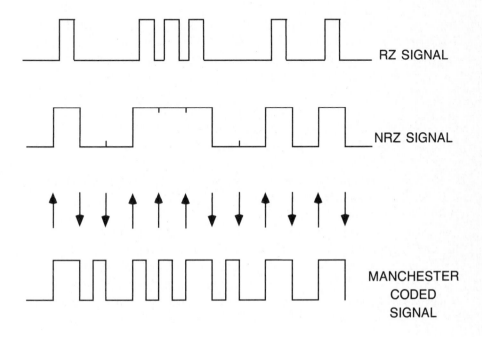

FIG. 4-20. Manchester coding.

Analog Signals

Baseband video is a good example of a troublesome analog signal type. A baseband NTSC video-plus-audio signal extends from a little over 4.5 MHz down to almost DC, with significant power components at 30 Hz due to the synchronization pulses. An amplifier with "flat" response over such a frequency range extending to 0 Hz is by no means a trivial design. Color video signals are extraordinarily sensitive to relatively minor distortions in amplitude and phase, with *differential phase* and *differential gain* errors greater than 1 degree and 1%, respectively, considered of significance.

Fortunately, the design work for such demanding applications has been performed repeatedly, and has been embodied in integrated circuit form. While it may be tempting to consider the problem to have been solved and no longer worthy of study, the techniques used are instructive.

Classical Techniques

Among the time-honored techniques for extending the bandwidth of an amplifier are collector-load compensation (for common emitter [source] configurations), emitter compensation (for common collector [drain] configurations) and, of course, DC coupling.

Since parasitics of significance at these frequencies are primarily capacitive, inductive elements chosen to sustain the load impedances at high frequencies are employed. At some frequency, however, the inductor will resonate with the parasitic capacitance, and the net impedance will peak and then decline at higher frequencies. In the meantime, the phase of the output signal will undergo drastic fluctuations. Controlling the net fluctuations requires elaborate networks and techniques such as stagger-tuning. Regrettably, integrated circuits do not include inductors in their component catalog, so that off-chip components or fundamentally different techniques must be resorted to.

The bandwidth-limiting *Miller Capacitance* effect can be dealt with effectively by utilizing the *Cascode* structure, dating back to vacuum-tube days (see Figure 4-21). It will be recalled that the collector of the bottom transistor of the cascode does not experience the gain-dependent (and inverted) voltage swings that otherwise effectively multiply the base-collector capacitance.

4.4 RADIATION EFFECTS

The circuits employed to drive the light-emitting devices and to amplify the signals from light-receiving devices are heir to all of the degradation effects typical of irradiated semiconductor circuits. The degradation is a function of the semiconductor technology applied, including the material system and circuit type as well as the radiation species and intensity.

Charge generated in the bulk semiconductor material is a major culprit which must be either avoided with an insulating substrate, dielectric isolation, or implantation of appropriate atoms, or overcome, once generated, by some other technique.

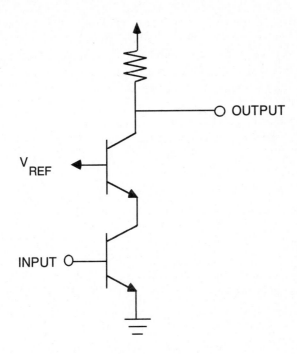

FIG. 4-21. The cascode circuit.

In general, bipolar devices are more sensitive than unipolar devices (though threshold voltages in the latter shift due to radiation-induced charges that arise and are trapped in the gate oxide [19]), and silicon devices are more sensitive than gallium arsenide types (Si is a semiconductor, while GaAs may be thought of as a semi-insulator). Radiation hardening techniques, ranging from processing variations to circuit tricks have proven effective [20].

Similarly, implementation details can be important. CMOS implemented as Silicon on Sapphire (SOS) is generally superior to bulk CMOS in transient radiation hardness (though they can be made comparable in total-dose radiation hardness; e.g., 500 krads); single-event upset immunity of 10^{-10} upsets/bit day is possible.

4.5 SUMMARY

The major distinguishing characteristics of the circuits used to drive light-launching devices are that they typically must be capable of both high-frequency and high-current. Favored structures include emitter-coupled logic configurations.

The circuits that serve ILDs must accommodate their peculiarities, including time and age sensitivity of the threshold current, and nonlinearity of the response. Threshold current shift may be detected via monitoring light output or (with some difficulty) electronic sensing.

Receiver circuits are primarily of two varieties: transimpedance and integrating. The transimpedance amplifier (the most-used) is essentially a short-feedback-path operational amplifier that converts an input current at high-impedance into an output voltage at low impedance. It has a wide dynamic range and often requires no equalization.

The integrating (or high-impedance) amplifier utilizes no feedback, and tends to integrate the incoming signal because of its low-pass characteristics. It has a poor dynamic range and requires equalization.

Linearization (needed primarily in association with transmission of analog signals) may be effected using feedback or feedforward techniques, or by judiciously choosing the frequency placement of the conveyed signals.

Postamplifiers are less complex because they deal with robust signals.

The radiation effects upon transmitting and receiving circuits are those typical of electronic circuitry in general.

EXERCISES

1. Design a common-base drive circuit for an LED. What advantages would such a drive have over alternatives?

2. Find the voltage across an ILD at threshold as a function of bandgap energy and device resistance. Calculate the voltage for $\lambda = 0.82$ and 1.3 μm for the resistance vanishingly small.

3. Consider the following approach to linearization: employ two LEDs driven by a phase splitter, each illuminating a fiber; at the receiving end, employ two p-i-ns whose outputs are combined with a differential amplifier. What could be expected of this configuration with respect to

improvement over a single LED? Could such a technique be extended to two wavelengths transmitted over the same fiber?

4. Diagram a feedforward nonlinearity correction circuit for a light-wave transmitter. Identify any implementation problems.

5. Invent a new circuit for driving LEDs or ILDs which is fundamentally different from those discussed.

6. What average incident power level does 1000 photons per "logical one" represent at a 100 Mbps signaling rate, assuming a wavelength of 1 μm?

7. Design a transimpedance amplifier useful over the frequencies of interest, and determine its performance with respect to gain and noise.

8. Perform the same exercise for an integrating amplifier.

9. What would the dynamic range of a typical preamplifier have to be in order for it to function as well when butted against the source as when a maximum length of fiber is inserted?

10. Identify some applications where timing recovery circuitry is unnecessary.

REFERENCES

[1] J. C. Daly, "Fiber Optic Intermodulation Distortion," *IEEE Transactions on Communications,* vol. COM-30, no. 8, August, 1982.

[2] F. Stremler, *Introduction to Communication Systems,* 2nd Ed., Addison-Wesley Publishing Co., 1982, p. 365.

[3] A. Albanese, "An Automatic Bias Control (ABC) Circuit for Injection Lasers," *Bell System Technical Journal*, vol. 57, 1978, pp. 1533-1544.

[4] P. T. Greiling, "The Future Impact of GaAs Digital IC's," *IEEE Journal on Special Topics*, vol. SAC-3, no. 2, March, 1985, pp. 384-393.

[5] H. Kroemer, "Heterostructure Bipolar Transistor and Integrated Circuits," *Proceedings of the IEEE,* vol. 70, 1982, pp. 13-25.

[6] R. G. Meyer, R. Eschenbach, W. M. Edgerly, Jr., "A Wide-Band Feedforward Amplifier," *IEEE Journal on Solid-State Circuits,* vol. SC-9, No. 6, December, 1974, pp. 422-428.

[7] A. Henschied, P. J. Barney, "CATV Applications of Feedforward Tech-
     niques," *IEEE Transactions on Cable Television,* vol. CATV-5, no. 2,
     April, 1980, pp. 80-85.

[8] C. Baack et al., "Analogue Optical Transmission of 26 TV Channels,"
     Electronics Letters, May 10, 1979, vol. 15, no. 10, pp.300-301.

[9] D. J. H. Maclean, *Broadband Feedback Amplifiers,* Research Studies
     Press, 1982, p. 181.

[10] Ibid, p. 17.

[11] Ibid, p. 187.

[12] B. L. Kasper et al., "An Optical Feedback Transimpedance Receiver for
     High Sensitivity and Wide Dynamic Range at Low Bit Rates," To be
     Published.

[13] J. E. Goel, "An Optical Repeater with High-Impedance Input
     Amplifier," *Bell System Technical Journal,* vol. 53, no. 4, April 1974, p.
     640.

[14] H. Kressel, *Topics in Applied Physics, Volume 39, Semiconductor Dev-
     ices for Optical Communication,* Springer Verlag, 1980, pp. 125-26.

[15] P. S. Henry, "Lightwave Primer," *IEEE Journal on Quantum Electron-
     ics,* vol. QE-21, 1985, pp. 1862-1879.

[16] T. Miki et al., *Technical Digest of the International Conference on
     Integrated Optics and Optical Fiber Communications,* Japan, 1977.

[17] M. Teich, "Avalanche Multiplication in Photodiodes," *NSF Workshop
     and Grantee-User Meeting on Optical Communications,* Ithica, New
     York, June 26, 1985.

[18] T. Li, "Advances in Lightwave Systems Research," *AT&T Technical
     Journal,* January/February, 1987, vol. 66, Issue 2, p. 9.

[19] J. Elster, "Survey of Semiconductor Foundries," *VLSI Systems Design,*
     August, 1986, p. 62.

[20] J. E. Gover and J. R. Srour, *Basic Radiation Effects in Nuclear Power
     Electronics Technology,* Sandia Report: SAND85-0776, May, 1985.

that these techniques are intermediate between analog and digital in the channel bandwidth they require.

Since the wide bandwidth of fiber encourages the transmission of video signals, it is useful to discuss the peculiarities of that signal type. Though the highest frequency in a standard NTSC color video signal including audio is about 4.5 MHz, implying that (applying the Nyquist criterion) a 9 MHz sampling rate would be sufficient, adequate recovery of phase (of great importance to accurate recovery of chrominance) without visible artifacts, in practice requires sampling at at least three times the color subcarrier rate ($\approx 3 \times 3.58$ MHz ≈ 10.7 Mega-samples per second). It is not unusual to sample at a yet higher multiple, and a standard of four times the color subcarrier (≈ 14.3 Mega-samples per second) has been proposed by the SMPTE (Society of Motion Picture and Television Engineers) [1].

PULSE AMPLITUDE MODULATION

Pulse amplitude modulation represents the sampled level with a proportionate pulse amplitude. It introduces no quantizing noise, but confers no noise immunity except in the case where the pulse is *strobed* so that any noise entering the channel during the non-strobed periods is eliminated. Any nonlinear influences in the channel will injure the information carried, so that relatively expensive, linear amplifiers are required. Successive impairments are irretrievable and cumulative, limiting the number of repeaters that can be used. It has essentially no perceived advantage for use with fiber systems or wideband networks except as a potential candidate for time division multiplexing. See Figure 5-1.

PULSE WIDTH MODULATION

Pulse Width Modulation (PWM) varies the width of a pulse in proportion to the amplitude of the sampled signal. No quantizing noise is suffered, and nonlinearities have no effect upon the contained information. Demodulation can be performed with a simple low-pass filter (the simplicity of the filter is dependent upon the nature of the signal conveyed; a video signal, for example, is relatively delicate, and requires a filter with carefully controlled phase characteristics). PWM can traverse logic gates unharmed, requiring, for example, only about 15 megapulse per second rates to convey standard NTSC TV signals. Further, it is asynchronous, obviating the need for retiming circuits. PWM can experience difficulty if transported over metallic facilities of any length to or from the photonic devices because such facilities will begin to perform the demodulation

	and Co-Channel Isolation Immune to Nonlinearities Can Traverse Gates Low Pulse Rate Simple Modulator and Demodulator Asynchronous No Quantizing Noise	Range Limited to a Few Spans
PPM	Similar to PWM but Superior in Jitter	Same as PWM
PCM	Best S/N Improvement and Co-Channel Isolation Moderate BER Tolerance Immune to Nonlinearities Can Traverse Gates Unlimited Range	Both Ends Complex ($) High Pulse Rates Quantizing Noise Requires Synchronization
ADM	Excellent Co-Channel Isolation Good BER Tolerance Immune to Nonlinearities Can Traverse Gates Moderate Pulse Rates Nonsynchronous Unlimited Range	Quantizing Noise Slope Overload Color and Picture Degradation in TV

TABLE 5-1: COMPARISON OF MODULATION METHODS.

5.3 PULSE ANALOG MODULATION

Pulse analog techniques include methods that vary a pulse's amplitude, width, position, or frequency in accordance with the modulating signal's sampled amplitude. The average pulse rate is the sampling frequency in all cases, so

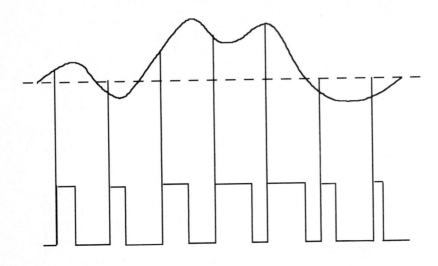

FIG. 5-2. Pulse width modulation.

applicable to short-range systems.

PULSE POSITION MODULATION

Pulse Position Modulation (PPM) varies the position of a constant-width pulse in proportion to the amplitude of the sampled signal. Its properties are similar to those of PWM, and in fact, a PPM signal can be derived from a PWM signal via differentiation and clipping.

PPM has some advantages for fiber transmission because the constant-width pulses place less of a constraint upon the light-launching and receiving devices and circuits [3]. For other techniques, the changing duty-cycle of the "on" portion of the pulse train can create distortions and promote jitter. See Figure 5-3.

5.4 FREQUENCY MODULATION

Frequency modulation is well-known as a broadcast radio means, and its attributes of high-fidelity and low noise in that application are known and appreciated by many. The high-fidelity characteristics of broadcast FM are in no sense an intrinsic property of the technique, however, but rather of the significant bandwidth expended. The low-noise characteristic, however, is intrinsic to FM

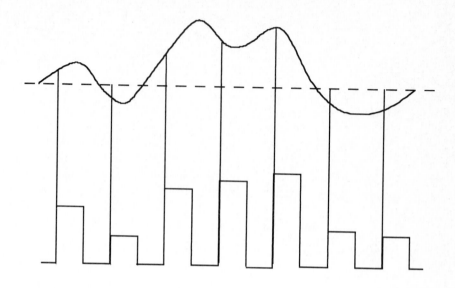

FIG. 5-1. Pulse Amplitude modulation.

process.

The signal-to-noise ratio improvement attributable to PWM is a function of the bandwidth allocated to convey the signal, which in turn is identifiable as a function of both the rise-time (and fall-time) of the pulses, the pulse rate, and the *deviation* of the pulse width. The improvement can be expressed as [2]:

$$(S/N)_o/(S/N)_i = \frac{1}{8} \left(\frac{B}{f_m}\right)^2 ,$$

where B is the bandwidth utilized, and f_m is the modulating frequency. See Figure 5-2.

A peculiar attribute of PWM is that crosstalk from another PWM signal is coherent in the most perceptually disturbing way possible for the case of video signals: sufficiently strong crosstalk is perceivable as a ghost image on the screen.

Because the information being transmitted is invested in the relative temporal position of a pulse edge, any jitter experienced constitutes noise. The contributions of jitter in a succession of photoelectronic conversions can accumulate unacceptable, uncorrectable net jitter. The technique is therefore primarily

CHAPTER 5

MODULATION TECHNIQUES

5.1 GENERALITIES

Modulation techniques are of interest relative to conveying information over fiber facilities because the properties of some techniques match the peculiarities of fiber transmission well, while others do not.

The term "modulation technique" here is intended to pertain to the form of processing performed upon the information before and after transmission over fiber facilities. If the fiber system facilities were transparent (no pun intended) to the process, such considerations would be unimportant. As will be seen, modulation considerations *are* important.

5.2 ANALOG MODULATION

A *literal analog signal* is a continuously variable physical parameter (e.g., voltage, current, photon flux) whose magnitude is in direct proportion to the magnitude of some information signal (e.g., the voltage amplitude of a baseband video signal) to be conveyed. In general, it possesses no intrinsic property to aid in its competition with noise, and can be taken as the base case in making comparisons with other techniques (though *companding*, a common adjunct to modulation, can be helpful).

The circuitry necessary to convert literal analog signals to optical equivalents and back again after reception is the simplest on a component-count basis. There are two primary disadvantages to the use of literal analog signals in fiber transmission systems: distortion resulting from nonlinearities introduced by the electrical to optical transducer, and susceptibility to noise introduced in the detector and in the receiver electronics.

The effects of the nonlinearities introduced may be made tolerable by limiting the bandwidth of the signals conveyed, limiting the extent of frequency multiplexing, spacing the frequency bands multiplexed to avoid major distortion products, wavelength division multiplexing, improving the linearity of the transmitter diodes, and/or using linearizing techniques in the transmitter and/or receiver circuits as discussed in the previous chapter.

The signal-to-noise ratio of a literal analog system can be made adequate by increasing the level of the launched light intensity range (while staying below amplitudes that would experience nonlinear fiber behavior), using lower loss fiber, making the transmission path shorter, providing shielding around the receiver, and using low-noise receiver circuitry (e.g., FET front ends or optimized high-impedance preamplifiers).

. Literal analog modulation, because of its simplicity and economy, might be a reasonable technique to consider for applications where the distances are such that the need for repeaters is rare, and the number of channels to be frequency multiplexed is small. It is, however, unsuitable for traversing digital broadband switching networks.

Table 5-1 contrasts the various techniques to be discussed. It will be seen that double-side-band AM provides a 3 dB signal-to-noise ratio advantage over baseband analog, essentially because it utilizes twice the bandwidth.

TYPE	ADVANTAGES	DISADVANTAGES
AM (SSB)	Simplest, Lowest Cost No Quantizing Noise	Requires Linear Amplifiers Little S/N Improvement Easily Injured by Noise, Crosstalk, Nonlinearities Cannot Traverse Gates
AM (DSB)	3 dB Improvement No Quantizing Noise	Same as SSB AM
FM	Excellent S/N Improvement and Co-Channel Isolation Immune to Nonlinearities Can Traverse Gates Moderate Pulse Rates No Quantizing Noise	Difficult to Deviate Large Percentages Linearly Range Limited to a Few Spans
PWM	Good S/N Improvement	Coherent Crosstalk

5.5 DIGITAL MODULATION

Digital modulation departs radically from other techniques in that information about a modulating signal amplitude is encoded in a form that does not degrade irretrievably unless the signal-to-noise ratio is very low. With other techniques, the signal-to-noise ratio monotonically (and irretrievably) decays, while digital techniques can promise virtually perfect signals essentially independent of distance. ("Virtual" perfection rather than absolute perfection is practically obtained because the technique itself introduces noise, and probabilistic considerations obviate a vanishing BER.)

Digital modulation techniques now include a number of variations upon the basic theme. Generally, the variations strive to reduce the bandwidth needs or the hardware necessary for transmitting digitally.

PULSE CODE MODULATION

Pulse Code Modulation (PCM) produces a succession of binary strings, each of whose weighted numerical value is representative of the amplitude of the signal at the time of sampling. Because the number of bits which may be employed is finite, only a discrete number of levels can be exactly represented, and the effect of the differences between the discrete levels chosen and the amplitude of the continuous signal to be represented is referred to as *quantizing noise*.

PCM can provide an almost arbitrarily high quality of transmission at a price in circuitry so long as the input signal-to-noise ratio is not excessively poor. For the typical 8-bit per sample, 3-times color subcarrier NTSC television signal sampling, some 88 Mbps are required, necessitating the use of very high-performance (and therefore relatively expensive) circuitry. As mentioned in Section 5.3, there has been some support for adopting a 4-times color subcarrier sampling rate, which would require a yet higher bit rate. See Figure 5-5.

The circuitry necessary to perform the A/D and D/A conversions is necessarily nimble, so that techniques such as *flash encoding* are often employed to perform the needed function. (Flash encoding employs a multiplicity of replicated circuits which perform the encoding in parallel, reducing the necessary speed of the circuits, but yielding the encoding in parallel, usually requiring ancillary parallel-to-serial converters; TRW was one of the first manufacturers to make such encoders available.) These *codecs* can be quite expensive, but the movement toward almost totally digital TV receiver signal processing should drive prices down precipitously via economies of production scale [7].

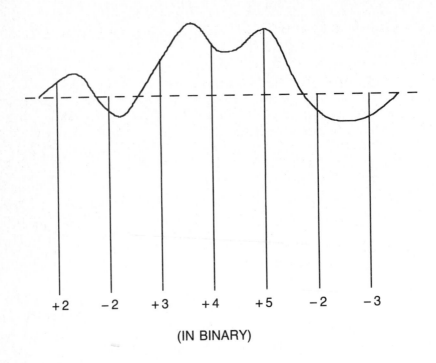

+2   −2   +3   +4   +5   −2   −3

(IN BINARY)

FIG. 5-5. Pulse code modulation.

It is necessary to provide synchronization means at the receiver in order to properly interpret the weights of the bits representing a sample. (This is of course essential as well when several channels are time multiplexed together.)

The signal-to-noise ratio can be given by [8]:

$$(S/N)_o \approx 1.3 \, e^{1/4 \, (S/N)_i} \qquad \text{for } (S/N)_i < 10 \text{ dB,}$$

$$\approx 2^{2n} \qquad \text{for } (S/N)_i > 15 \text{ dB,}$$

where n is the number of bits utilized to represent each sample.

PCM is becoming a powerfully pervasive transmission technique, so much so that it tends to be chosen over all others. Its popularity and standardization have promoted the manufacture of integrated circuits to serve it, in such volume that the prices have begun to rival those for analog circuits. Having the signal in digital form also allows storage in memory media and the use of digital processing techniques, including filtering, conference bridging, echo cancellation, etc.

DELTA AND ADAPTIVE DELTA MODULATION

Delta Modulation (DM) transmits only a differential signal. For example, a "one" may be transmitted per sample when the slope of the modulating signal is positive, and a "zero" when it is negative [9, 10]. Because so little information is transmitted per sample, it must be transmitted at higher rates in order to fashion an acceptable replica at the receiver.

The circuitry necessary is relatively simple, making this technique particularly attractive, though the sampling rate must in general be considerably higher than for PCM. For signals such as television, however, which include very high-rate-of-change pulses for synchronizing, DM tends to fall short, suffering from a malady called "slope overload." See Figure 5-6.

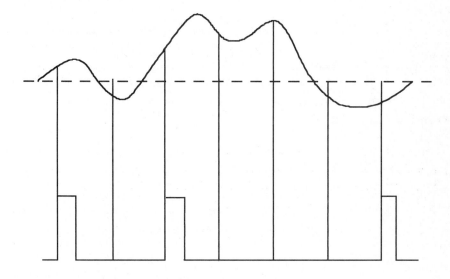

FIG. 5-6. Delta modulation.

Adaptive Delta Modulation (ADM) attempts to overcome slope overload by changing the increment represented by a bit when a succession of like bits have been transmitted [11]. Additional circuitry is required, reducing the advantages of this technique.

Both of these techniques provide signals that can traverse logic gates, and are of a form highly resistant to nonlinear distortion. They are also basically nonsynchronous in that they need not be retimed at the receiving end, and the pulse frequency can vary considerably without harm. Quantizing noise is generated, and for video signals there can be color and picture degradation taking on the forms of hue impairment, "busy" edges, and granularity.

The signal-to-noise ratios provided by these techniques are highly dependent upon the details of the implementation.

5.6 DIGITAL CARRIER

With conventional, nonoptical systems, it is often necessary to impress a digital signal upon a carrier whose nominal frequency is significantly greater than twice the bit rate of the modulating signal. Three distinct cases can be identified, ASK (Amplitude-Shift Keying), FSK (Frequency-Shift Keying), and PSK (Phase-Shift Keying).

These techniques are of interest in optical transmission because the (light) carrier is of the order of 10^{14} Hz, and when coherent techniques can be applied, great improvement in range and/or bandwidth usage become possible.

In the following, the three techniques will be only briefly outlined, relegating to a later chapter discussion of coherent detection techniques for light transmission. A more thorough discussion of the topics can be found e.g., in Peebles [11].

AMPLITUDE-SHIFT KEYING

Amplitude-Shift Keying (often called OOK for On-Off Keying), literally extinguishes the carrier during a space, and turns it on again for a mark. The pulse envelope may be rectangular or cosine or raised cosine.

FREQUENCY-SHIFT KEYING

Frequency-Shift Keying alters the frequency of the carrier between marks and spaces. It can be viewed as the superposition of two ASK signals with differing carrier frequencies. When duty cycles are taken into account, it can be shown that the error probability of this technique is identical to that of ASK.

PHASE-SHIFT KEYING

Phase-Shift Keying inverts the carrier between marks and spaces. It can be viewed as the superposition of two ASK signals shifted in time relative to each other. This technique enjoys an error probability that is 3 dB better than ASK or FSK.

5.7 CHOICE OF MODULATION TECHNIQUE

The requirements of a particular application usually significantly narrow the range of choices among the modulation schemes. Low budget, undemanding applications may use simple amplitude modulation, while applications for which performance requirements are stringent and the budget is generous would tend to pick PCM. Intermediate applications may be satisfied by one of the pulse-analog approaches or some variation of the differential PCM technique.

As will be discussed in a later chapter, PSK has properties particularly favorable to coherent detection techniques, and becomes the modulation method of choice.

5.8 SUMMARY

Several modulation techniques are appropriate to fiber optics system transmission. Analog signals may be carried in applications where the distortion due to transmitting devices is not critical or can be compensated for. Pulse analog modulation schemes such as PWM and PPM are viable, as is square-wave FM. All digital formats such as PCM, DM, and ADM are similarly appropriate.

EXERCISES

1. If square-wave FM is utilized, one popular demodulation scheme is termed "pulse-counting." It differentiates, then rectifies the signal, producing pulses whose temporal density reflects the amplitude of the modulating signal, and demodulation becomes easy. One scheme clips off the negative pulses before final demodulation, while another full-wave rectifies first, providing a higher signal amplitude. Suggest a reason why the latter method might be less attractive for a signal transmitted over fiber optics.

2. Given a maximum bandwidth, which modulation scheme will provide the best S/N ratio? (Make reasonable assumptions with regard to modulation indices, bits per sample, etc.)

3. Perform a first-order design of modulators and demodulators for the modulation techniques named, compare the complexity, and comment on likely relative costs:

   a. Pulse-Width Modulation

   b. Frequency Modulation

   c. Pulse-Code Modulation.

4. Consider the feasibility of the following dual-modulation scheme: combine pulse-amplitude and pulse-width modulation by varying the amplitude of each pulse in proportion to the sampled amplitude of one signal and the width of each pulse in proportion to the sampled amplitude of another. Ideally, the two modulations are in quadrature and should be separable. Suggest a separation technique and discuss practical problems that might surface in an implementation.

5. Reconcile the signal-to-noise expressions for PCM for the case when $(S/N)_i$ is between 10 and 15 dB.

6. What features of a video signal render it particularly difficult for differential PCM to accurately convey? Suggest a remedy for such difficulties.

7. Design a circuit for the production of PPM from PWM via differentiation and clipping.

8. Apply Carson's Rule to determine the bandwidth necessary to convey a video signal FM modulating a 50 MHz signal with a 10 MHz deviation. How many FM-modulated video channels could be conveyed on a typical fiber if the light-launching device linearity were not limiting?

REFERENCES

[1] E.g., C. P. Ginsburg, "Report of the SMPTE Digital Television Study Group," *SMPTE Technical Conference*, Detroit, 1976.

[2] W. D. Gregg, *Analog and Digital Communication*, Wiley, 1977, p. 76.

[3] S. D. Personick, *Optical Fiber Transmission Systems*, Plenum Press, 1981, p. 157.

[4] W. D. Stanley, *Electronic Communications Systems*, Reston, 1982, pp. 197-98.

[5] A. B. Carlson, *Communication Systems*, 2nd Ed., McGraw-Hill, 1975, pp. 236-7.

[6] E.g., P. F. Panter, *Modulation, Noise, and Spectral Analysis*, McGraw-Hill Book Company, 1965, p. 740.

[7] E.g., C. Schepers, "A Digital CTV Chassis Concept," *1983 IEEE International Conference on Consumer Electronics: Digest*, pp. 88-89.

[8] Panter, pp. 662-673.

[9] D. Slepian, "On Delta Modulation," *Bell System Technical Journal*, vol. 51, no. 10, December, 1972, p. 2101.

[10] M. Schwartz, *Information Transmission, Modulation, and Noise*, McGraw-Hill, 1980, pp. 147-49.

[11] E.g., P. Z. Peebles, *Communication System Principles*, Addison-Wesley, 1976, p. 393ff.

CHAPTER 6

SYSTEMS

6.1 SYSTEMS CONSIDERATIONS

The effective application of fiber optics equipment requires consideration of the entire system: the modulation technique, the transmitter, the light launcher, the fiber, its length and characteristics, the connectors, the splices, the receiving detector, the preamplifier, the post-amplifier, the signal-to-noise or BER requirements, etc. A wide and growing range of systems have been designed and deployed, with very favorable results to date. Some telephone administrations (e.g., the British Post Office [1]), have declared policy decisions to abandon the use of copper facilities in favor of fiber optics for interoffice applications; such policies are likely to spread, and when fiber begins to appear in a significant volume in the telephone loop plant, the effect upon copper prices may become noticeable. (The domestic US loop plant has already seen fiber in the form of feeder cable for the AT&T Subscriber Loop Carrier remote line concentrator [2].)

This chapter will address the pertinent considerations important to system design, and will present examples of practical systems.

6.2 SIGNAL-TO-NOISE RATIO CONSIDERATIONS

The signal-to-noise ratio requirements for a system are a function of the modulation technique chosen. Analog techniques require signal-to-noise ratios of the order of 30-40 dB (electrical) [3], while BERs of the order of 10^{-9} for PCM may require only 20 dB (electrical) signal-to-noise ratios [4] (depending upon the data rate).

6.3 FIBER BANDWIDTH AND ATTENUATION

The bandwidth and attenuation characteristics of the fiber must be adequate for the intended application. In practice, segments of an overall route may be comprised of fibers with differing characteristics. Designers must be cognizant of such considerations, and ideally, should have prior data from installed or at least on-spool measurements available to them. Alternatively, *Time-Domain Reflectometry* may be employed to take attenuation data and detect any anomalies on the spot.

It is important to note that the performance of a system cannot be inferred from the characteristics of any one constituent component. It is of interest, for example, to determine the maximum bandwidth (or repeater spacing) for a particular assemblage of components. Figure 6-1 illustrates the nature of such an analysis. Because of BER requirements, the level of power launched into the fiber, and the receiver sensitivity, there will exist an attenuation limit as shown, which droops due to the receiver sensitivity dependence on bit rate; similarly, a dispersion limit exists.

Note from considerations in Chap. 2 that the dispersion limit is a function of wavelength, and of the spectral width of the source. In particular, systems driven by LEDs tend to be dispersion limited (the dispersion-limit is shifted to the left), while ILD-driven systems tend to be attenuation limited.

6.4 LOSS BUDGET

Postponing bandwidth considerations for a time, it is convenient to approach the design of a system via evaluating a *loss budget* which spells out the loss contribution of each component. Alternatively, the maximum capabilities of a given set of devices, circuits, fiber, splicing technique, and connector type can be evaluated.

LAUNCHED POWER

The power available to be launched into the fiber is a reasonable starting point for a budget calculation (not unlike gross income in a monetary budget analysis).

The power successfully launched is a function of the light source, the optics, and the fiber size and numerical aperture. Since fiber core diameters vary in size from about 5 μm (single-mode) to about 200 μm (multi-mode), and in numerical aperture from about 0.1 to 0.5, and the size of the emitting device and degree of collimation vary widely, the fraction of generated light actually launched into the fiber varies from about 0.01 to 0.5 (though higher values have been demonstrated). Optics can aid to some extent.

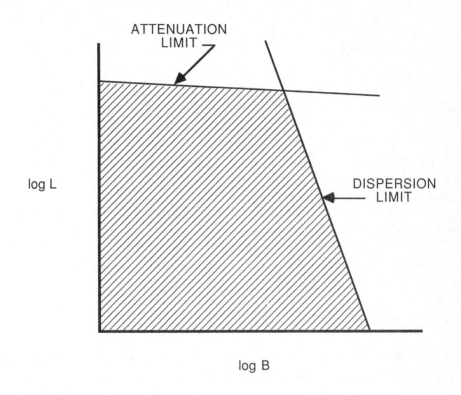

FIG. 6-1. Attenuation and dispersion limits.

FIBER

The attenuation exacted by the fiber depends upon its composition, its refractive index grading (if any), the operational wavelength (or wavelengths), its temperature, its radiation exposure history, and, of course, its length.

Typical values range from 5 dB per km to 0.2 dB per km for high-silica glass, but may be as high as 100s of dB per km for plastic. Actual fiber may prove somewhat nonuniform from sample to sample, and a statistical analysis is advisable. A "worst-case" analysis is possible, but may lead to gross over-design and increased cost.

SPLICES

The virtually inevitable splices also take their toll of the system budget.

Splice losses range from about 0.1 dB for very expertly performed fusion joining down to a mean of 0.03 dB for an actively-aligned butt splice [5] utilizing an index matching gel. Poor and very bad splices can be very much worse, ranging toward infinite attenuation.

CONNECTORS

Connectors usually profit from manufacture under controlled conditions (in the factory or laboratory environment) and tend to display losses comparable to the better butt splices. Field-applied connectors may be significantly poorer in quality because of the non-ideal ambient conditions and generally less well-trained splicers. For military field use, where cleanliness, experience, and the time for careful handling cannot be depended upon, special structures may be resorted to which are relatively undemanding (e.g., large-area connectors tolerant of "dirty" surfaces).

DETECTORS

Coupling losses associated with the detector are usually quite small. Unlike the problem of launching generated light at the transmitting end of a fiber, the light issuing from the receiving end can be collected with high efficiency by a large detector. Nonreflective coatings reduce the light lost at the interface, and most of the photons which have survived the trip will be effective in producing hole-electron pairs which are collectible and contribute to the signal current.

RECEIVER SENSITIVITY

After all of the lossy influences have reduced the available received power, the receiver must be able to interpret the incoming information with the required degree of accuracy in the presence of corrupting noise. As has been mentioned, in the case of digital information, a BER is usually specified. Typical receiver sensitivities are somewhat poorer than those depicted in Figure 4-18.

AGING

The light-launching devices tend to display reduced output with age. For LEDs the process is not usually catastrophic, but is gradual. For ILDs, it is sometimes fulminating, and can go to rapid decline. Gradual degradation in either device type usually stems from the growth of "dark spots" which are regions that cease to emit light. Crystallographic defects have been identified as one of the major culprits. Catastrophic failure in ILDs may be the result of facet damage due to overdriving. (Interestingly, in applications where the driving circuitry for the light-launching device has safe-guards against overdrive, the ILD may be viewed as a gradually degrading device. Note also that drive circuitry compensates for aging effects over a significant range of decline.)

Light receiving devices are under the stress of reverse bias, particularly APDs, which may have quite high voltages applied. Devices which have shown no hint of trouble sometimes spontaneously break down.

Butt splices may exhibit increased loss with time due to accumulation of contaminants upon or between surfaces, and index matching fluid (if used) may degrade.

The fiber itself has aging mechanisms. Hydrogen intrusion has been implicated in reducing the transparency of fiber over time [6]. The degree to which this phenomenon may prove a problem in long-term practice remains to be demonstrated.

Radiation, both natural and man-made, if intense or over long duration, can degrade the transparency of fiber. A profound degradation of short term after intense radiation may be followed by a partial recovery. The choice of dopants materially affects radiation performance [7].

DISPERSION PENALTY

Because ILDs (when used) seldom exactly match the zero-dispersion wavelength taken advantage of in high-performance systems, the consequent intersymbol interference and mode-partition noise require additional input power to overcome their effects. Limited bandwidth fiber utilized near its limits also contributes bandwidth-induced signal degradation. These effects are lumped under the term *dispersion penalty*.

The dispersion penalty has the effect of shifting the diagonal lines in Fig. 4-18 (Receiver Sensitivities) upward and to the right, requiring more photons per bit to maintain the same BER, or degrading the BER.

The penalty is difficult to measure, being a function of the transmitter, fiber and receiver, and measurement requires sophisticated instrumentation. For a typical system it has been found to be expressible in the form (in dB)

$$P_d = k \left[\frac{Bit\ Rate}{B} \right]^2 , \qquad\qquad 6.1$$

where k is a value to be determined by measurement, and B is the bandwidth of the section of fiber measured with the intended application source type [8].

The dispersion penalty is usually below a dB or two in a power budget, and it will be lumped into the aging allowance for computational purposes in the following (but it should be kept in mind that it bears no relation to aging).

THE BUDGET

A typical loss budget may be made up taking into account all of the above items:

$$P_E - L_C - l\ \alpha_f - n\ L_S - m\ L_m - A \geqslant S_R/\eta, \qquad 6.2$$

where

P_E = Emitted Power
L_C = Diode-to-Fiber Coupling Loss
l = Span Length
α_f = Attentuation per Unit Length
n = Number of Splices
L_S = Average Splice Loss
m = Number of Connectors
L_m = Average Connector Loss
A = Aging Margin
η = Detector Quantum Efficiency
S_R = Minimum Receiver Sensitivity

EXAMPLE

Suppose that

P_E = 12 dBm
L_C = 12 dB
l = 10 km
α_f = 3 dB/km
n = 5 Splices
L_S = 0.2 dB
m = 4 Connectors
L_m = 1 dB/Connector

$$\eta \approx 1$$
$$A = 5 \text{ dB}.$$

Then the minimum receiver sensitivity $S_R \approx - 40$ dBm. This value is readily attainable for a BER of 10^{-9} and bit rates $\leqslant$ about 100 Mbps [9]. Figure 6-2 displays the corresponding power-level diagram. (Note that in this figure and subsequent renderings of power-level diagrams, the pigtail fiber leads assumed to exist in association with the light sources and the detectors are depicted out of proportion to their actual length [which would typically be a few inches]; their contribution to loss is negligible as is reflected in the diagrams.)

The above treatment of loss budget is intended to be illustrative. Actual analyses may be carried out somewhat differently. A manufacturer might, for example, simply quote a *facility loss budget*, which is the difference between a power budget and a system allowance which may include end of life, eye degradations, modal variation, dispersion loss, optical feedback loss, transmitter wavelength variation, etc.

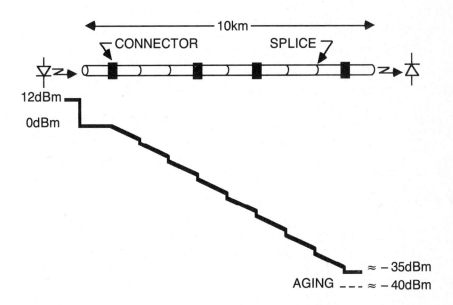

FIG. 6-2. Power-level diagram: example.

6.5 BANDWIDTH CONSIDERATIONS

BANDWIDTH BUDGET

The system bandwidth can be budgeted across the individual subsystem components as well.

It will be recalled that the overall risetime t_s of a system can be calculated from

$$t_s{}^2 = \sum_{i=1}^{n} t_i{}^2 , \qquad\qquad 6.3$$

where the t_i's are the individual risetimes of the system elements. For a fiber system, these elements are the transmitter, the light source, the fiber span, the detecting device, and the receiver complex (preamplifier, filter, equalizer, and discriminator).

It should be noted that a transmission system will, in general, comprise a concatenation of a multiplicity of fiber segments, each of which may be unique in its bandwidth characteristics.

In practice, except for very ambitious designs, the bandwidths of the individual system components can readily be made (via judicious choice) to exceed the needs of the system application by a sufficient margin to attain the needed capability.

How may a "judicious choice" be made? It is possible to tentatively choose the source - fiber - receiver system in the following fashion: as a rule of thumb:

> Short, LAN-like systems tend to employ short-λ (e.g., 0.82 μm) LEDs - fat GI Fiber - silicon-detector receivers of modest sensitivity;

> Medium length, inter-office trunk-like systems tend to employ long-λ (e.g., 1.31 μm) multi-longitudinal mode ILDs - GI or SM fiber - InGaAsP detector receivers of moderate sensitivity;

> Future long haul, upper hierarchy toll-office trunk-like systems may employ long-λ (e.g., 1.55 μm) single-longitudinal mode ILDs - dispersion-shifted SM fiber - InGaAsP or germanium detector receivers of high sensitivity.

> (Very short, low-speed systems may employ visible-λ LEDs -

plastic or plastic-clad glass fiber - silicon detector, low sensitivity receivers.)

Having categorized an application, manufacturers' data sheets can be scanned to gather sufficient data to calculate bandwidth and determine whether the preliminary component choice is adequate. In particular, from considerations discussed in Chapter 2, the zero-dispersion wavelength, the dispersion slope, and the rms width of the source spectrum can be used to calculate the dispersion for single-mode fiber.

For multimode fiber, if the source is a short-λ (e.g., 850 nm) LED, then material dispersion dominates, if it is a long-λ (e.g., 1300 nm) LED, then both modal and material dispersion contribute. If the source is a long-λ ILD, then modal considerations totally dominate.

It can be shown that rms dispersion contributions are related by [10]

$$\sigma_{total}^2 = \sigma_{material}^2 + \sigma_{modal}^2 . \qquad 6.4$$

Since, assuming a Gaussian pulse shape, the bandwidth is related to the rms dispersion by

$$B = \frac{1}{8\sigma} , \qquad 6.5$$

it follows that

$$\frac{1}{B_{total}^2} = \frac{1}{B_{material}^2} + \frac{1}{B_{modal}^2} . \qquad 6.6$$

It should be emphasized that the bandwidth referred to in a multi-mode fiber specification sheet is the "modal" bandwidth measured with a laser source caused to excite all propagation modes (but with a line width rendering material dispersion negligible). To calculate the bandwidth when an LED source is to be used, it is necessary to know the dispersion at the wavelength of interest and the source spectral width. Thus, while a fiber illuminated by an LED at 1.3 μm might display a bandwidth-distance product of 1 GHz-km, that product might shrink to only 80 MHz-km at 0.85 μm because operation is so far from the dispersion zero. It is pointlessly expensive to insist upon fiber with a high "factory" bandwidth when operation with a short-λ LED is contemplated.

MULTIMODE FIBER BANDWIDTH UNDER CONCATENATION

Length Dependence

In general, multimode fiber bandwidth displays a length dependence given by

$$B = \frac{k}{L^{\gamma_{cb}}},$$

where L is the length, and γ_{cb} is the *cut-back bandwidth length dependence parameter* [11] applicable for spliced lengths of identical fiber. γ_{cb} may range from 0.5 to 1.0, depending upon the fiber and the wavelength of operation (it approaches 1.0 when the operation of the fiber is far from the fiber's maximum bandwidth, and 0.5 when operated near that value). Typically it is about 0.9. k is the bandwidth-distance product of the fiber measured using a spectrally narrow source, e.g., a laser. The bandwidth-distance product is typically measured by the manufacturer while the fiber is spooled prior to cabling. Additional factors may be introduced to account for packaging, installation and length-scaling effects [12].

For a multiplicity of multimode fibers with differing index profiles concatenated via splicing, the following relation applies [13]:

$$B^{-\frac{1}{\gamma_{cat}}} = \sum_i B_i^{-\frac{1}{\gamma_{cat}}},$$

where γ_{cat} is the concatenation bandwidth scaling factor. As is true for γ_{cb}, it ranges between 0.5 and 1.0, but is typically around 0.7. (The 0.5 figure corresponds to complete mode mixing, and 1.0 to no mode mixing.) This expression can provide good estimates of the bandwidth of concatenated fiber.

6.6 SIZE CONSIDERATIONS

In some applications, the size of the transmission medium is very important. In downtown Manhattan, New York, for example, there are places where the space available for augmenting channel capacity is bounded above by the street level, below by the subway roof, and in the intervening space by a variety of water, steam, powerline, and sewage pipes, so that only existing telephonic conduit with limited and already 100% utilized cross-section is available. The ability to extract a portion of the metallic medium and substitute the much smaller-sized (even with cabling) and much wider-bandwidth fiber, is a nearly miraculous answer to the problem of expanding capability.

Similarly, anywhere size is important (e.g., avionics applications), fiber is a potential answer.

6.7 WEIGHT CONSIDERATIONS

Though most applications benefit little from the weight savings of fiber compared to metallic equivalents, in aircraft, this parameter is critically important. Coupled with its noise immunity, the weight savings is causing fiber to be considered more and more for avionics duty. The AV8-B Harrier II is a prime example of such an application.

6.8 OTHER CONSIDERATIONS

It is important to scan the manufacturer's data sheet for information on macrobending loss and mode-field diameter (usually in the range of 8.8 to 10 nm). The macrobending loss figure determines the ability of the fiber to be used with low loss in tight quarters such as splice cases, where the fiber is often provided with spare length which is repeatedly folded with loose bends. The mode-field diameter is a measure of the fiber sensitivity to microbending, and its propensity to splice loss. Minimum macrobending loss and mode-field diameters are clearly desirable.

Sometimes a *minimum repair length* is specified (e.g., 10-20 m). Splice joints (e.g., at a repair splice) may generate higher than fundamental modes which may be out of phase with the fundamental at the next splice, generating modal noise. The minimum repair length is that length which will have caused these higher-order modes (which are beyond cutoff) to have attenuated to a negligible value.

6.9 HIGH-FREQUENCY REQUIREMENTS

When high-frequency signals (ranging from some 100 Mbps up to a Gbps and beyond) must be accommodated, additional constraints are placed upon all elements of the system. (Often the term *broadband* is associated with such signals and the equipment dealing with them even when the bandwidth is relatively narrow at the high frequencies used.)

HIGH-FREQUENCY LIGHT SOURCES

LEDs have primarily been employed at the lower range of frequencies, but, e.g., 220 Mbps devices are commercially available, and experimental operation in the GHz region has been reported [14, 15]. Because of the doping changes necessary for the higher frequency operation, however, the power output of the devices is reduced roughly in inverse proportion.

ILDs are practically essential at present for frequencies above 200-300 Mbps, and have capabilities well beyond 1 Gbps.

HIGH-FREQUENCY DRIVING ELECTRONICS

The driving electronics, apart from the added complication of serving ILDs, become somewhat exotic at very high frequencies. It may be necessary to resort to discrete components (e.g., microwave transistors), strip-line conductors, differential drive, and careful shielding to deliver the necessary massive current changes at such high rates. The high dissipation at elevated frequencies must also be dealt with.

HIGH-FREQUENCY FIBER

The overall bandwidth of the fiber must, of course, be adequate for the task. Graded index fiber of good quality will afford about 500 MHz-km capability (2 GHz-km fiber has been produced). Single-mode fiber is required for higher-performance applications, and a choice of wavelength at or near a dispersion zero is called for (this requirement may be relaxed somewhat if a single-longitudinal mode laser is employed).

HIGH-FREQUENCY DETECTORS

Detectors can be fashioned which are capable of functioning at virtually any frequency at which a light-launching device can be modulated. Attention to the dimensions of the absorbing layer, trading off Responsivity against speed (e.g., by reducing the device area), or carefully choosing a wavelength and corresponding detector material system, can usually provide an adequate device.

HIGH-FREQUENCY RECEIVER ELECTRONICS

As usual, the receiver electronics are the most-pressed in extreme applications. As the bandwidth of the receiver is expanded, noise becomes more of a problem. Careful design, adequate shielding, and attention to power-supply decoupling can aid. In particular, RC filters as shown in Figure 6-3 are often effective. Ideally, two capacitors are used, one an electrolytic of significant value for low frequency decoupling, and in parallel, a mica or ceramic, noninductive type for high frequency decoupling.

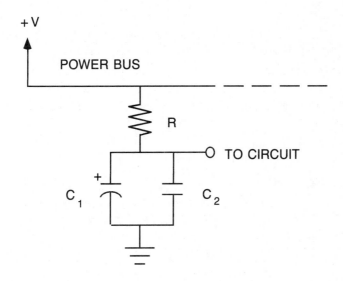

FIG. 6-3. Power-supply decoupling.

The resistor functions to permit the raw supply and the filtered supply to entertain differing instantaneous potentials, and the larger capacitor is chosen to provide transient currents with little voltage droop. Sometimes distinct decoupling approaches are used on the semiconductor chip, the chip carrier, the circuit card, and at the connector.

THE MODULATION TECHNIQUE

The choice of modulation technique can be important as well in broadband applications. The tradeoff between signal-to-noise ratio and bandwidth is well known, and can be exploited in certain situations. Since the bandwidth of fiber optic systems can be so large (larger than needed in many applications), the excess may be utilized to advantage in other dimensions, such as improving the BER.

6.10 COSTS

Typical costs at this writing and estimates of future values (expressed in 1988 dollars) for some of the significant constituents are as follows:

FIBER

Graded-Index

Good-quality graded-index fiber costs about $0.30/m cabled in reasonable cross-sections (number of fibers per cable). $0.10/m is a likely minimum future cost. (Caution: some manufacturers quote prices for uncabled fiber without calling attention to that fact; cabling can be a very significant cost component, particularly in small cross-section cables.)

Single-Mode

Single-mode fiber is somewhat less in cost per cable meter than graded-index fiber at this writing. Eventually, as its manufacturing volume increases, it may cost perhaps one-half as much as graded-index fiber.

LIGHT-LAUNCHERS

LEDs

LEDs, depending upon wavelength, frequency response, power output, and configuration (surface- or edge-emitting) cost in the range of $0.25 to $100.00 (for reasonable volumes; quantity-of-one costs can be significantly higher). Eventually, the best devices may settle in the range of $1.00 to $10.00.

ILDs

ILDs, again depending upon wavelength, frequency response and output power, range in price from $10.00 to $1,000.00 (and higher, if rated for ultra reliability or are at an uncommon wavelength). Eventual prices may drop to the $5.00 to $100.00 range.

6.11 SYSTEM LOSS-BUDGET EXAMPLES

It is often helpful to consider a number of examples spanning the range of real systems. The approach introduced earlier will be used for uniformity and to ease comparison. (In the following, it will be assumed that η is approximately unity.)

A VERY SHORT-RANGE DATA LINK

Very short-range links are usually the least demanding of the applications, and can therefore employ the least expensive and lowest quality components. Occasionally, very demanding short-range applications arise, however, requiring the fastest, best, and most expensive componentry.

Consider the case of a 30 m link which is to run at 10 MHz. Clearly, plastic fiber is a candidate for this application. With a bandwidth of 10 MHz-km, that aspect will present no problem. Attenuation may present more of a challenge depending upon the rest of the system. The NA can be comparatively large, which will aid in coupling. The wavelength must be chosen to be in the high-transparency range for plastic fiber (typically 0.7 μm).

EXAMPLE:

$$P_E = -10 \text{ dBm}$$
$$L_C = 10 \text{ dB}$$
$$l = 30 \text{ m}$$
$$\alpha_f = 100 \text{ dB/km}$$
$$n = 0 \text{ Splices}$$
$$L_S = \text{(Not Applicable)}$$
$$m = 2 \text{ Connectors}$$
$$L_m = 3 \text{ dB}$$
$$A = 10 \text{ dB}.$$

Required receiver sensitivity calculates to -39 dBm.
Figure 6-4 depicts the corresponding power-level diagram.

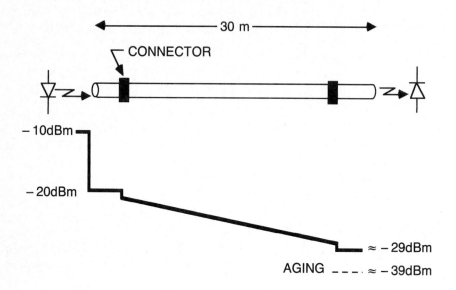

FIG. 6-4. Power-level diagram: very short-range link.

Most plastics undergo physical changes with age and with exposure to hostile environments including extreme heat and radiation. There does not yet appear to be sufficient data to support or condemn plastic fiber with regard to aging at moderate temperatures, but extreme temperature applications should be approached with caution.

A SHORT-RANGE APPLICATION

When the span length is increased to 1 km, presently available plastic fiber becomes prohibitively lossy, and glass becomes the material of choice (though plastic-coated silica might be a candidate for moderate environmental conditions). The dispersion over such a distance would allow very-good-quality step-index fiber to be employed at frequencies up to 100 MHz.

EXAMPLE:

$$P_E = -5 \text{ dBm}$$
$$L_C = 20 \text{ dB}$$
$$l = 1 \text{ km}$$
$$\alpha_f = 5 \text{ dB/km}$$
$$n = 1 \text{ Splice}$$
$$L_S = 1 \text{ dB}$$
$$m = 4 \text{ Connectors}$$
$$L_m = 1 \text{ dB/Connector}$$
$$A = 5 \text{ dB.}$$

Receiver sensitivity calculates to -40 dBm.
See Figure 6-5.

A MEDIUM-RANGE APPLICATION

A medium-range application might have an unrepeatered span length of about 10 km and run at 100 Mbps. It would employ LEDs and p-i-ns, and graded-index fiber.

EXAMPLE:

$$P_E = 0 \text{ dBm}$$
$$L_C = 15 \text{ dB}$$
$$l = 10 \text{ km}$$
$$\alpha_f = 1 \text{ dB/km}$$
$$n = 4 \text{ Splices}$$
$$L_S = 0.5 \text{ dB}$$
$$m = 6 \text{ Connectors}$$
$$L_m = 0.75 \text{ dB/Connector}$$
$$A = 5 \text{ dB.}$$

The receiver sensitivity calculates to -36.5 dBm.
See Figure 6-6.

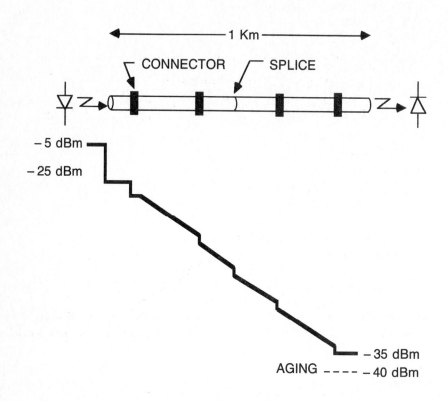

FIG. 6-5. Power-level diagram: short-range link.

A LONG-RANGE APPLICATION

Long-range applications include span-lengths of 40 km and perhaps 420 Mbps data rates. ILDs and single-mode fiber would be required.

EXAMPLE:

$$P_E = 10 \text{ dBm}$$
$$L_C = 15 \text{ dB}$$
$$l = 40 \text{ km}$$
$$\alpha_f = 0.3 \text{ dB/km}$$
$$n = 10 \text{ Splices}$$
$$L_S = 0.1 \text{ dB}$$
$$m = 4 \text{ Connectors}$$

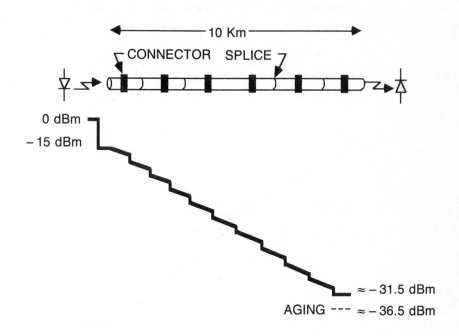

FIG. 6-6. Power-level diagram: medium-range link.

L_m = 0.2 dB/Connector
A = 10 dB.

The receiver sensitivity calculates to -28.8 dBm.
See Figure 6-7.

A UNIQUE EXAMPLE: THE TRANSATLANTIC SUBMARINE CABLE

TAT-8, the first Trans-Atlantic fiber optic system, operational in June of 1988, is comprised of 3,146 nautical miles of cable deployed by AT&T, 279 nautical miles deployed by UK-based Standard Telephone & Cables Ltd., and 167 nautical miles deployed by Submarcom, a French group [16]. The system transmits 560 Mbps on 2, single-mode fiber pairs. Tuckerton, New Jersey, Widemouth, England, and Penmarch, France are joined by the cable (see Figure 6-8).

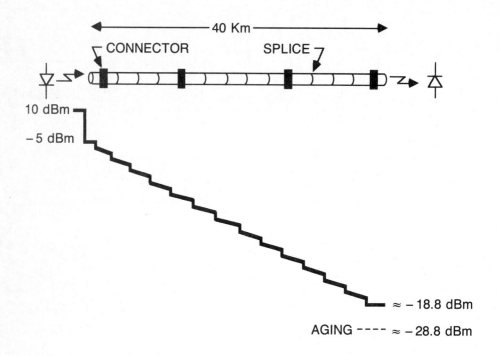

FIG. 6-7. Power-level diagram: long-range link.

It should be recognized that this system is a dramatic departure from the traditional, analog coaxial cable systems, which have virtually reached the limits of their capabilities. New techniques have been pressed into service, and the results will affect thinking for years to come in such applications.

Repeater spacing is 30 nautical miles, six times the spacing for conventional cable. The repeaters are powered from shore, with 2,500 volts at 1.5 amperes being provided. Operation at depths as great as 6,000 meters will be required.

Indium-gallium-arsenide-phosphide laser diode transmitters are employed, operating at 1.3 μm. P-(almost)i-n (actually p+n-n+) diode detectors are used in the receivers (intrinsic regions are difficult to produce with this material system). Operating temperatures are in the range of 10° C to 60° C.

Three fiber pairs are employed, one each to England and France, and one spare. DCMS (Digital Circuit Multiplication System) is employed to increase the nominal 8,000, 64-kbps voice channels to 40,000 equivalent channels. DCMS combines digital speech interpolation and low bit-rate voice encoding

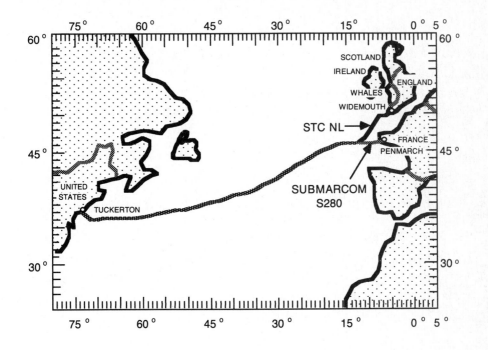

FIG. 6-8. TAT-8: The transatlantic optical cable.

[17].

All splices employ fusion techniques for low loss and high tensile strength.
A possible set of parameters follows:

$$P_E = 5 \text{ dBm}$$
$$L_c = 5 \text{ dB}$$
$$l = 60 \text{ km}$$
$$\alpha_f = 0.4 \text{ dB/km}$$
$$n = 6 \text{ Splices}$$
$$L_s = 0.15 \text{ dB}$$
$$m = 0 \text{ Connectors}$$
$$L_m = (\text{Not Applicable})$$
$$A = 5 \text{ dB}$$

The receiver sensitivity calculates to about -30 dBm. See Figure 6-9.

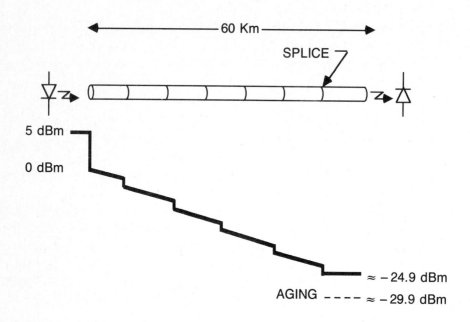

FIG. 6-9. Power-level diagram: transatlantic submarine link.

For completeness, it should be pointed out that a similar facility, the Hawaii/Transpac3 (or TPC-3) fiber optic system, will span the Pacific with service starting in 1989 (see Figure 6-10). A 2,250 nautical mile segment will connect Point Arena, California with Makaha, Oahu, Hawaii. From Hawaii, the cable will continue 2,900 nautical miles to a branching unit which will split the cable into two routes extending some 850 nautical miles to Agana, Guam, and about 1,200 nautical miles to Boso, Japan. Operation will be at 280 Mbps. Some 250 repeaters will be spaced at 30 nautical mile intervals on the sea bottom.

The Australian Overseas Telecommunications Commission has announced plans to deploy a fiber optics undersea cable called Tasman 2 which will link Australia to Asia and North America in the mid-1990s [18]. The first phase will connect Australia to New Zealand in 1991. An application has also been filed with the FCC to land and operate a fiber-optics cable between Washington state and Japan in 1989.

TAT-9, involving AT&T and 24 other companies, is planned for service in 1991. It will interconnect the US, Canada, the United Kingdom, France, and Spain, and provide fiber access to a number of Mediterranean countries. It will function over two fiber pairs at 565 Mbps, with an operating wavelength of 1.5 μm.

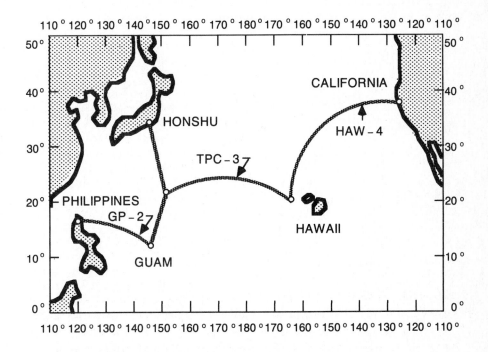

FIG. 6-10. TPC-3: The transpacific optical cable.

6.12 TERRESTRIAL SYSTEM EXAMPLES

It is useful to trace a progression of systems offered by one major manufacturer (AT&T) because it mirrors the advancing technology and the demand for ever-greater bandwidth. The last example in the preceding section is a submarine system; the following, in contrast, are terrestrial systems.

Table 6-1 from [19] illustrates seven years of progress in telecommunications-oriented terrestrial system offerings.

SYSTEM	YEAR	BIT RATE	CAPACITY (64 kbps chs.)
FT3	1980	45 MBPS	672
FT3C	1983	90 MBPS	1,344
FTX-180	1984	180 MBPS	2,688
FT3C + FTX180	1984	270 MBPS	4,032
FT Series G 417	1985	417 MBPS	6,048
FT Series G 417	1986	834 MBPS	12,096
FT Series G 1.7	1987	1668 MBPS	24,192

TABLE 6-1 AT&T's FT Series Transmission Systems

The FT3 system was a commercial offering following the "Chicago Experiment" begun in 1977 [20]. The bit rate is a North American trunking standard. It makes use of 0.82 μm ILDs, graded-index 50/125 μm fiber, and silicon APDs. The primary application is for interoffice trunks over distances < 7 km.

The FT3C system was first utilized in sections of the "Northeast Corridor," a route stretching from Boston to Richmond via New York, Philadelphia, Baltimore and Washington, DC. It uses the same componentry as the FT3 system at a higher rate.

The FTX-180 system utilizes 1.3 μm lasers with repeater spacing doubled due to lowered loss at longer wavelength. It too was applied in the Northeast Corridor as an *overbuild* on the FT3C routes to yield a 270 Mbps net rate.

The FT3 Series G 417 system utilizes single-mode fiber at 1.3 μm. It was first applied in a Philadelphia-to-Chicago route. This system was augmented via wavelength-division multiplexing at two wavelengths (1.3 and 1.55 μm) to provide 834 Mbps transmission.

The FT3 Series G 1.7 system continues the progression by redoubling the maximum rate [21]. Still higher rate systems are in the offing [22].

Subscriber loop carrier (SLC) systems have been developed that utilize fiber in the *feeder* portion of the route, between the telephone central office and the remote terminal from which the distribution loop cable emanates. Early systems used LEDs and multimode fiber, but the latest systems use either ILDs or edge-emitting LEDs to drive single-mode fiber [23].

6.13 LOCAL AREA NETWORKS

Local Area Networks (LANs) are a distinct variety of transmission system because they typically are utilized over very short distances, and may employ taps for picking off and injecting signals into the bit stream. They come in a variety of architectures, bit rates, signaling protocols, etc. Typically they tend to utilize large-diameter core (e.g., 62.5 μm) graded-index fiber for ease of coupling, and LEDs and p-i-ns. These systems are emerging in such variety as to defy brief discussion. The reader is referred to the current literature.

6.14 EXPERIMENTAL SYSTEMS

A number of experimental systems have been put into service throughout the world, primarily in telephone environments, and therefore by telephone interests.

The advent of optical fiber immediately sparked interest in the possibility of its application in the telephone loop plant (the portion of the telephone transmission plant that interfaces directly with the customer) because the majority of the transmission medium is utilized in that portion of the plant. The Integrated Services Digital Network (ISDN), an evolving international standard generated as a CCITT (an international standards-making body) recommendation, has been expanded to include broadband services (BISDN), and has encouraged study and experimentation utilizing fiber as the loop transmission medium.

Because fiber systems remain significantly more expensive than 26 Gauge twisted pair cable facilities (the loop standard in much of the world), it cannot "prove-in" (yet) economically unless it can provide additional services and command the attendant additional revenues. An obvious service that fiber can convey is entertainment video; customers have demonstrated a willingness to pay for such service by subscribing to cable TV. Data services, perhaps more immediately applicable to business, are believed to be of growing utility to residential customers as well. These services may include standard data rates for computer communication as well as telemetry capability for meter reading, fire and intruder alarms, etc.

Once the equipment necessary to provide TV to the customer over fiber has been deployed, the incremental cost of other services is small. A system that can provide voice, video and data services to the customer at a reasonable cost may therefore be the ultimate answer, and telephone administrations in many parts of the world are sufficiently convinced to have deployed experimental systems to test hardware, customer acceptance of services, etc., in anticipation of eventual reductions in the cost of the necessary technology.

As will be seen, the approaches taken by the several countries or administrations differ in detail, but the basic philosophies are uniform: utilize optical fiber and ancillary technology to bring to the customer voice, data and video in as economical a manner as possible.

Among the problems all such systems must wrestle with are: choice of mechanism for powering the subscriber equipment (from the Central Office via additional conductors or from local mains at the subscriber); if the latter is chosen, determination of whether to provide backup batteries on the subscriber's premises in case of power outage; whether to use multiple fibers or a single fiber bidirectionally; whether to use frequency division or wavelength division; etc.

JAPAN

Japanese interests have been extremely active in the field of fiber optics from its inception, and have been very diligently pursuing the possibility of loop-plant application.

HI-OVIS

The HI-OVIS field trial began in 1978 and continued until 1983 [24]. It provided broadcast and local two-way TV, utilized step-index fiber, 0.83 μm LEDs, silicon p-i-ns, and served 168 customers in Higashi-Ikoma, Nara, a suburb of Osaka, Japan. Two fibers per customer were supplied, one for downstream and the other for upstream communication; both fibers were within the same cable, and the longest subscriber loop was 1.16 km in length. A 200 bps program selection and control data stream was impressed above the upstream video signal at 6.6 MHz.

The fiber utilized plastic cladding over a glass core, and was produced via the VAD process; the operating temperature range was -20° C to $+60^o$ C (at 90° C, the cladding would begin to degrade). The core diameter was 150 μm, and the overall diameter was 350 μm. The effective numerical aperture was 0.25. The fiber bandwidth-distance product was 20 MHz-km, and its loss was less than 10 dB/km at the wavelength used. The LEDs produced an average optical output power of 1 mW, and the coupling loss was 8 dB. The nonlinearity of the device was countered by using diode compensation circuitry and emphasis-deemphasis techniques. The p-i-n diodes displayed a quantum efficiency of 0.8 at 8 volts bias. The receiver, which utilized both FET and bipolar devices, was designed to provide a S/N ratio of greater than 46 dB with an input of -30 dBm. Fusion splices yielded losses of less than 0.5 dB, and connector losses averaged less than 1 dB. [25]

A variety of programming was provided, including retransmitted television, studio-produced live programs, fixed time and on-demand fare, still-picture programs such as news, and premium channels.

This experiment was the earliest large-scale trial of fiber in the loop plant, and influenced many other efforts.

YOKOSUKA

The Yokosuka field trial ran from 1980 through 1981 [26]. It was considerably broader in scope than HI-OVIS, providing Teleconferencing, High-speed FAX, and Digital PBX services to business customers, and broadcast TV (including High-Definition capability), Digital Data, and "TV Telephone" to residential customers; these services encompassed all aspects of the ISDN (Integrated Services Digital Network) as defined at that time. One graded-index fiber was used per customer, and wavelength-division multiplexing of up to four wavelengths was utilized. Both laser diodes (at 0.80, 0.83, 0.86, and 0.89 μm) for business customers, and LEDs (at 1.15 and 1.3 μm) for residential customers were used, as were Silicon or Germanium APDs (depending on wavelength).

TOKYO: INS

Nippon Telephone and Telegraph has constructed a model of a system called INS (Information Network System) in the Mitaka office in Tokyo which offers video conferencing, video transmission, and high-speed, high-definition facsimile which serves 300 customers. The system utilizes multiplexing of four different wavelengths on the same fiber. This is a model of a planned nationwide system called HBN (High-Speed, Broadband Network) [27].

FRANCE

The French government has embraced the concept of applying fiber optics to telephony and video to the point of establishing, in July of 1982, a law (the Mexandeau law) which establishes the general rules and structures to govern the distribution of broadcast service in France [28]. Upon passage of the law, the French PTT immediately launched bids to construct several local area networks, and systems to serve up to one million customers have been proposed.

BIARRITZ

The trial at Biarritz is intended to serve some 3,000 customers (the first phase [now implemented] serves 1,500, and the central equipment and infrastructures permit expansion to 5,000), offering cable TV, digital voice, Videotex, and videotelephone services [29, 30]. Plans have been made to connect more than one million French households to a nationwide optical cable network. It is estimated that more than three million homes will have been connected to fiber optics systems by 1989, including the towns of Montpellier, Rennes, and parts of Paris [31]. Mantes, Sevres, St. Cloud, Suresnes, Evry, and Toulon will also be cabled for optical videocommunication networks [32]

Frequency Modulation is used to convey TV programming (two channels), videotelephone, and two-channel high-fidelity sound; speech for the videotelephone, and signaling and data are in PCM format.

Graded-index fiber with a 50 μm core diameter, a 125 μm diameter at the cladding, 0.20 numerical aperture, 3.5 dB/km attenuation, and a bandwidth better than 375 MHz-km is utilized. ILDs at 0.85 μm and p-i-ns are used for downstream transmission, and LEDs and APDs are used for upstream transmission. The maximum subscriber line is 2.2 km long. The lasers launch a minimum power level of 0 dBm into the fiber; they utilize a rear-facet monitoring diode. The LEDs launch -20 dBm. The receivers employ transimpedance amplifiers.

8,450 km of fiber and 20,000 optical connector plugs were used in the first phase. The feeders are 70-fiber cables, and 10-fiber cables are used for distribution. The drop cables comprise one or two fibers. (Feeders are the transmission means between the central office or head-end and a branch point or distribution center; distribution cables span between a branch point and a drop terminal; drops are the cables between a drop terminal and a customer terminal.)

GERMANY

The Deutsche Bundespost of the Federal Republic of Germany plans to have 540,000 kilometers of glass fiber in the local network and 281,000 kilometers in the toll network by the end of 1990 [33]. It plans to connect some one million videotelephone subscribers by 1995, and has formulated a trial to stimulate activity by several equipment suppliers [34].

BIGFON

Seven cities were selected for the BIGFON field trial: Berlin, Hanover, Hamburg, Dusseldorf, Stuttgart, Nuremberg and Munich. Six German manufacturers are providing at least one "optical island" per city which serves from 28 to 48 subscribers, 25% of whom have videotelephone terminals [35, 36]. Telephone, telex, data, TV, and "picture telephone" services are planned. The precise nature of the implementations differ as offered by the several manufacturers.

Video signals are digitally encoded and transported at about 140 Mbps.

320 subscribers are receiving telephone service, and 68 are receiving television and other services as well. The installation of the entire network was completed in the autumn of 1986.

BELGIUM

The Belgian Telecommunications Administration is considering at least two proposals for an integrated broadband local network demonstration project. One of these proposes to offer to the subscriber 1.1 Gbps downstream and 70 Mbps upstream over single-mode fiber [37].

SWEDEN

The Swedish Telecommunications Administration and Ericsson have jointly been testing a fiber video system between Farsta and Skarpnack (near Stockholm) since 1984 [38]. Households are provided with several video channels via fiber using wavelength-division multiplexing [39].

DENMARK

Studies are being conducted at Jutland Telephone Co. and the Technical University of Denmark toward an all-digital, all-optical national network [40].

SWITZERLAND

The Swiss PTT has begun activity toward providing broadband services.

MARSENS

In Marsens, Canton of Fribourg, a pilot project was set up to provide subscribers with TV and radio programs and to experiment with broadband return channels, including video conferencing and videotelephone [41].

ITALY

A broadband fiber system for TeleMedicine has been operative in Turin since 1982, and plans are being made for a more general system with emphasis upon videoteleconferencing and Videotex services [42]. An experiment is planned for installation in Rome, with conducting cables for the early phase, and fiber considered for later phases [43].

For the latter applications, efforts are being expended toward the use of a single fiber to the business or home, utilizing bidirectional and multiwavelength approaches.

THE NETHERLANDS

The Netherlands is performing several large coaxial cable broadband experiments, but a fiber experiment, called DIVAC, is ongoing at Geldrop [44].

DIVAC

Philips Co. in cooperation with the Dutch PTT and the Universities of Delft and Eindhoven, is conducting the DIVAC experiment at Geldrop. Various broadband services are provided for the office and residence.

ENGLAND

The Minister of Information Technology of the British Government has declared plans to accelerate the build-up of a nationwide electronic grid of broadband cable, and new legislation supporting such programs is scheduled. To support technology in this regard, a field-trial at Milton-Keynes has been on-going.

MILTON-KEYNES

The Milton-Keynes experiment, called FIBREVISION, became operational in 1982 [45, 46]. Eighteen new homes were connected to a head-end some three km away, and commercial television sets were employed. Pulse Frequency Modulation was employed throughout in order to reduce the complexity of the system.

Video signals are transmitted from the head-end to a remote node on individual fibers utilizing edge-emitting LEDs, APDs, and fusion-spliced fiber with 4 dB/km loss and a 400 MHz-km bandwidth. The remote node transmits over two fibers with 10 dB/km loss and 200 MHz-km bandwidth to the customers over distances up to 200 m using p-i-n diode detectors. The strength members of the two fibers are copper-coated and are used to convey signaling information upstream from the customer premises.

CANADA

The Department of Communications of the Canadian Federal Government has encouraged experimentation in broadband applications of fiber, including the trials at Yorkville and Elie.

YORKVILLE

A fiber optic system serving 40 customers with telephone, data and video services was installed in 1978 in downtown Toronto [47]. It employed analog transmission, graded-index fiber, and LEDs and p-i-ns in a star topology. The maximum fiber length was 1.4 km.

ELIE

A trial begun in Elie, Manitoba in 1981 and completed in 1984 offered telephone, Telidon (a version of Videotex) information, FM radio, and TV services to 150 customers in a relatively remote area [48]. Most of the customers were served with two active fibers, but a few (25) were served with a single fiber utilizing two wavelengths and directional couplers to provide bidirectionality. The fibers were 50 μm core, 125 μm cladding in size. Telephone, CATV (both fixed-time and on-demand), and data services were provided. Several modulation schemes were employed: FDM and FSK downstream and PCM-TDM upstream.

Most subscriber loops employed LEDs at 0.84 μm, while longer loops (> 3 km) used ILDs at 0.93 μm. The detectors were APDs.

A star-star topology was employed, utilizing a feeder link to a Remote Distribution Center which in turn distributed links to individual customers.

Though the trial has been completed, the system is to be maintained as an operational unit and testbed until 1990 [49].

THE UNITED STATES

HUNTER'S CREEK AND HEATHROW

Southern Bell, one of the BellSouth (a regional holding company) operating companies, has a field trial under installation by AT&T in Hunter's Creek, near Orlando, Florida [50, 51]. Initially, some 300 homes are to be served with 45 Mbps video over two, multimode fibers dispatched from a controlled environment vault terminating 36 active singlemode fibers from a head-end. A later phase is planned to provide integrated voice, data and video services.

In Heathrow, Florida, Southern Bell has installed (1988) another trial to initially serve 500 homes with voice, data and video services over single-mode fiber [52]. In addition to advanced (ISDN) voice data and CATV services, security monitoring, remote meter reading, and home energy management capabilities will be offered.

LEAWOOD

Work is underway by Southwestern Bell to install an AT&T single-mode fiber system to serve 50 to 100 homes in Leawood, Missouri, a suburb of Kansas City [53]. The trial is scheduled to begin in October, 1988 and will provide conventional telephone service over fibers to the individual homes.

6.15 RADIATION EFFECTS

Previous chapters have enumerated the degradation effects upon individual constituents of a fiber optic system. If an entire system is likely to be subjected to irradiation, the combined effects of the several degradations must be evaluated. In particular, a modified loss budget can be formulated which can aid in determining which components will be most critical in a given application.

Similarly, from a system viewpoint, the possibility of single-event upset and/or of transient (but lengthy in electronic time) complete disruption of the system should be considered. Ideally, the system will be made tolerant to such possibilities.

6.16 SUMMARY

System design requires attention to all aspects of a fiber-optics system. Of particular interest are bandwidth and attenuation characteristics.

A loss budget can be set up to allocate loss across a system.

Size and weight considerations can greatly influence system choices.

Much of the world is planning and, in many cases implementing, trial fiber systems to provide broadband services to the end user.

EXERCISES

1. Optoisolators, which may transmit light over distances of less than a mm, are normally only available with relatively low frequency performance, e.g., 20 Mbps, which is a small fraction of the bandwidth of a typical fiber link. What considerations could be limiting the performance of such devices? Design a high-performance optoisolator (e.g., 200 Mbps) using

fiber-optic system components, and formulate a loss budget.

2. Design a system which must transmit a 60 MHz analog signal for a distance of up to 300 feet using inexpensive fiber, end photonics and electronics. Draw a power-level diagram.

3. Design a system which must transmit a 200 Mbps signal for a distance of 4 km. Draw a power-level diagram.

4. Design a system which must transmit a 1 Gbps signal for 100 km. Draw a power-level diagram.

5. Design a system which must transmit a 500 Mbps signal for 12,000 miles (half-way round the world). Draw a power-level diagram.

6. Design a system with the highest imaginable channel capacity on one fiber over the longest imaginable span.

7. Invent a method for "sparing" ILDs in submarine cable repeaters. I.e., suggest a method for automatically replacing a failed ILD with a viable one without having to raise the repeater to the surface (such a technique is planned for TAT-8).

8. Critique the field trials described.

9. Describe the philosophies and implementation choices you would make if called upon to design a fiber system to provide Broadband ISDN services to residential and business customers.

REFERENCES

[1] P. W. Lines and D. Millington, "Optical Fiber Systems for the British Telecom Trunk and Junction Networks," *National Telecommunications,* IEEE, 1980, p. 46.4.3.

[2] P. P. Bohn et al., "Bringing Lightwave Technology to the Loop," *Bell Laboratories Record,* April, 1983, pp. 6-10.

[3] S. E. Miller and A. G. Chynoweth, *Optical Fiber Telecommunications,* Academic Press, 1979, p. 671.

[4] P. K. Cheo, *Fiber Optics Devices and Systems,* Prentice-Hall, 1985, p. 266.

[5]   C. M. Miller, *Optical Fiber Splices and Connectors*, Marcel Dekker, Inc., 1986, pp. 342-345.

[6]   M. Miyamoto et al., "Effects of Hydrogen on Long-Term Reliability of Optical Fiber Cable," *Conference on Optical Fiber Communications*, San Diego, February, 1985, p. 46.

[7]   E. J. Friebele, P. C. Schultz, M. E. Gingerich and L. M. Hayden, "Effect of B, P, and OH on the Radiation Response of GE-Doped Silica-Core Fiber-Optic Waveguides," *Digest, Topical Meeting on Optical Fiber Communication*, March 1979, Washington, DC., p. 36.

[8]   J. J. Refi, "Fiber Bandwidth and its Relation to System Design," *FOC/LAN 86*, Orlando, Florida, October, 1986, pp. 251-257.

[9]   H. Kressel, *Topics in Applied Physics, vol. 39*, Springer-Verlag, 1980, p.139.

[10] R. Olshansky and D. B. Keck, "Pulse Broadening in Graded-Index Optical Fibers," *Applied Optics*, vol. 15, February, 1976, pp. 483-491.

[11] T. Ito and K. Nakagawa, "Transmission Experiments in the 1.2-1.6 Micrometer Wavelength Region Using Graded-Index Optical Fiber Cables," *Fiber and Integrated Optics*, vol. 3, no. 1, 1980, p. 7.

[12] J. J. Refi.

[13] P. R. Reitz, "Characterization of Concatenated Multimode Optical Fibers with Time Domain Measurement Techniques," *Proceedings of the Conference on Precision E-M Measurements*, Boulder, Colorado, 1982, p. L-14.

[14] H. Nomura, "Very Fast LEDs Offer Response for Optical Local Area Networks," *Journal of Electronic Engineering*, vol. 22, no. 217, January, 1985, pp. 34-38.

[15] *Electronics Week*, January 21, 1985, p. 25.

[16] P. K. Runge and P. R. Trischitta, "The SL Undersea Lightwave System," *IEEE Journal on Selected Areas in Communication*, vol. SAC-2, no. 6, November, 1984, pp. 784-793.

[17] J. Piasetsky and R. Murphy, "DCM Systems to Expand TAT-8 Circuit Capacity," *Telephony*, August 4, 1986, pp. 70-74.

[18] "Plans for Two Trans-pacific Cables Announced," *Telephony*, June 16, 1986.

[19] R. J. Sanferrare, "Terrestrial Lightwave Systems," *AT&T Technical Journal*, January/February, 1987, vol. 66, issue 1, pp. 99-107.

[20] H. Kressel, *Topics in Applied Physics*, vol. 39, 1980, Springer Verlag, pp. 259-284.

[21] R. W. Lay, "AT&T Plans 1.7-Gbit Fiber Link Between Philadelphia and Chicago," *Electronic Engineering Times*, April 21, 1986, p. 32.

[22] H. Reiner, "Integrated Circuits for Broad-Band Communications Systems," *IEEE Journal on Selected Areas in Communications*, vol. SAC-4, no. 4, July, 1986, pp. 480-7.

[23] T. D Nantz and W. J. Shenk, "Lightguide Applications in the Loop," *AT&T Technical Journal*, vol. 66, issue 1, January/February, 1987, pp. 108-118.

[24] K. Sakurai and K. Asatani, "A Review of Broad-Band Fiber System Activity in Japan," *IEEE Journal on Selected Areas in Communication*, vol. SAC-1, no. 3, April 1983, pp 428-435.

[25] S. Takeuchi, "An Optical Fiber Cable Video System," *Fiber and Integrated Optics*, Crane, Russell, & Co., Inc. vol. 3, no. 1, 1980.

[26] K. Asatani, K. Nosu, T. Matsumoto, and K. Yanagiomota, "A Field Trial of Fiber Optic Subscriber Loops in Yokosuka," *ICC '81 Conference*, vol. 3, Paper 48.1.1-5.

[27] S. Harashima and H. Kimura, "High-Speed and Broad-Band Communication Systems in Japan," *IEEE Transactions on Selected Areas in Communications*, vol. SAC-4, no. 4, July, 1986, pp. 565-572.

[28] Loi no 82-652 du 29 Juillet 1982 "Sur la Communication Audiovisuelle," *Journal Official de la Republique Francaise*, July 30, 1982, p. 2431.

[29] "Telecommunications par Fibres Optiques," Comite Consultatif International Telegraphique et Telephonique, *Internal Review of CNET*, Lannion, France, 1983.

[30] R. T. Gallegher, "Cabled City Forms French Prototype," *Electronics*, June 14, 1984, pp. 88-92.

[31] *Fiber Optics News*, July 14, 1986.

[32] H. Seguin, "Introduction of Optical Broad-Band Networks in France," *IEEE Journal on Selected Areas in Communications*, vol. SAC-4, no. 4, July, 1986, pp. 573-578.

[33] Deutsche Bundespost, "Mittelfristiges Programm fuer den Ausbau der technischen Kommunikationssysteme," (Medium Range Program for the Evolution of the Communications Systems)

[34] "An Ever-Changing World Market," *Photonics Spectra,* January, 1985, pp. 62-63.

[35] J. Kanzow, "BIGFON: Preparation for the Use of Optical Fiber Technology in the Local Network for the Deutsche Bundespost," *IEEE Journal on Selected Areas in Communications,* vol. SAC-1, no. 3, April, 1983, pp. 436-439.

[36] R. T. Kneisel, "Broad-Band Communication Systems in Germany," *IEEE Journal on Selected Areas in Communications,"* vol. SAC-4, no. 4, July, 1986, pp. 579-588.

[37] R. David et al., "Integrated Broad-Band Communication Systems in Belgium," *IEEE Transactions on Selected Areas in Communications*, vol. SAC-4, no. 4, July, 1986, pp. 596-604.

[38] B. Stierlof and D. Westin, "Fiber Optics Improves Broadband Transmission Quality," *Telephony*, June 1, 1987, pp. 38-39.

[39] A. Hansson and S. Jacobson, "Wavelength Division Multiplexing for Fibre-Optic Subscriber Lines," *Ericsson Review*, vol. 62, no. 4, 1985, pp. 170-174.

[40] A. Monclavo and F. Tosco, "European Field Trials and Early Applications in Telephony," *IEEE Journal on Selected Areas in Communication*, vol. SAC-1, April, 1983, pp. 398-403.

[41] J. Buetikofer and U. Stettler, "Pilot Network for Wideband Communications at Marsens, Switzerland," *Tech Mitt PTT,* vol. 61, no. 12, 1983, pp. 414-420.

[42] C. Marchesi et al., "Analogue Transmission of Video Signals and Applications to Telemedicine," *CSELT Technical Report*, vol. XII, no. 2, April 1984, pp. 189-191.

[43] A. Fausone, A. Monclavo, and F. Tosco, "A Strategy for Broad-Band Network Introduction in Italy," *IEEE Transactions on Selected Areas in Communications*, vol. SAC-4, no. 4, July, 1986, pp. 605-611.

[44] J. Van der Heijden, "Divac: A Dutch Integrated Services Glass Fibre Experimentation System and Demonstration Set-Up," *Proceedings of the IEEE International Conference on Communications*, 1984, pp. 1145-1149.

[45] J. R. Fox, D. L. Fordham, R. Wood, and D. J. Ahern, "Initial Experience with the Milton Keynes Optical Fiber Cable TV," *IEEE Transactions on Communications,* vol. COM-30, no. 9, September, 1982 pp. 2155-2162.

[46] R. Wood and D. Moore, "From Fibrevision to the MultiStar Wideband Network," *Proceedings of SPIE,* vol. 403, Paris, May, 1983, pp. 64-70.

[47] R. L. Gallawa, "U.S. and Canada: Initial Results Reported," *IEEE Spectrum*, vol. 16, no. 10, October, 1979, pp. 72-73.

[48] K. Chang and E. Hara, "Fiber-Optic Broad-Band Integrated Distribution - Elie and Beyond," *IEEE Journal on Selected Areas in Communications,* vol. SAC-1, April, 1983, pp. 439-44.

[49] E. H. Hara, "Integrated Broad-Band Service - The Intelligent Building Strategy," *IEEE Journal on Selected Areas in Communications*, vol. SAC-4, no. 4, July, 1986, p. 619.

[50] S. Ubis, "Southern Bell Trial Sends Fibers to the Home," *Telephony*, September 8, 1986, p. 25.

[51] J. Wolfe, "Three Telcos Map Fiber Net Plans That May Pose Near-Term Challenge," *Cablevision*, May 18, 1987, pp. 50-56.

[52] M. Warr, "Fiber to Home in 'City of the Future'," *Telephony*, September 21, 1987, p. 13.

[53] C. Wilson, "Southwestern Bell, AT&T Take Fiber to the Home in First Trial," *Telephony*, vol. 213, no. 9, August 31, 1987, p. 10.

CHAPTER 7

FUTURES

7.1 FUTURES

Few fields have experienced the rate of change of technology enjoyed by fiber optics to date. A number of the "future" technologies that will be discussed in this chapter may therefore reach fruition in very short order.

7.2 ULTRA LOW-LOSS FIBERS

Before discussing the quest for ever lower-loss fiber, it is worth considering where the point of diminishing returns may lie. Even if the fiber loss were made zero, the loss over a span can be considerable due to other influences. For example, there are practical limitations to how long a length of fiber can be before a splice will be necessary. If the loss of a splice is comparable to or greater than the loss in the span length between splices, then it is at least as important to devote resources to reducing splice losses as it is to pursue reducing fiber losses.

Another consideration in the pursuit of ultra-low loss fibers is damage due to background radiation. The TAT-8 transatlantic cable, for example, is expected to experience 100 mrad/year of ionizing radiation which would have a negligible effect upon attenuation of the silica fiber over a 25 year span [1]. Such a loss, however, if it were of similar magnitude for an ultra-low loss fiber, could become a much more significant fraction of the loss budget.

Consider the dependencies of the length of a span:

$$l = \frac{10}{\alpha} \log \frac{P_{in}}{P_{out}},$$

where α is the attenuation in dB.

As Midwinter has observed [2] the span length can be increased by inP_{in}, but fiber nonlinearities constitute a limit. The necessary P

reduced, but this has practical limits due to noise; the activity in coherent detection techniques described in a later section pursues this goal. Lastly, the attenuation of the fiber can be reduced, and this is the goal of longer wavelength applications with silica glass, and of the pursuit of other materials.

There exist materials capable of guiding light which have extremely low-loss characteristics, but are unsuitable for practical application (at least as perceived at this time), e.g., solubility in water [3]. Means may nonetheless be found for applying these materials in practical systems. Of particular interest at present are *Halide Glasses* which employ combinations of certain of the halides (e.g., Cl, Fl, Br) in glass-yielding compounds [4], particularly the heavy metal fluoride glasses.

The halide glasses generally have their minimum attenuation at longer wavelengths than silica glasses, and their commercial application would require the availability of sources and detectors capable of operation at such wavelengths. It will be noted that photons at longer wavelengths are less energetic, so that a greater flux is required to maintain a given optical power level. Thus, for example, 5 times as many photons per unit time would have to be received at a wavelength of 5 μm as at one μm. On the other hand, pulses made up of larger numbers of photons should lead to less statistical jitter at the receiver. Further, the power level corresponding to the quantum limit will be proportionately reduced.

The mechanism leading to lower loss in the indicated new materials fibers is a reduction of the level of one end of the inevitable attenuation "bathtub." The bathtub is bounded at the lower wavelength end by Rayleigh scattering brought about by random disorder in the glass comprising the fiber. The boundary at the long-wavelength end is due to oxide-bond stretching vibration absorption mechanisms. The minimum attenuation will occur at the wavelength of intersection of these two components, so that pushing the bathtub ends apart tends to reduce the minimum attenuation [5].

7.3 WAVELENGTH DIVISION MULTIPLEXING

ultiplexing is being applied commercially (e.g., the
tween Boston and Washington, D.C. [6] which origi-
ee wavelengths nominally at 0.825, 0.875, and 1.3 μm,
bps for a net of 270 Mbps), but was implemented using
5 μm, and a 180 Mbps link at 1.3 μm. However, only a
rceived potential capacity in number of wavelengths has

been employed practically (10 wavelength transmission over single-mode fiber has been demonstrated experimentally, utilizing multiplexing equipment capable of 22 wavelengths [7]). It has been estimated [8] that it will be possible to accommodate some fifty distinct wavelengths in a single fiber. For many wavelengths to be launched, transmitted, and effectively separated requires extraordinarily narrow-spectrum sources that are very stable with temperature (otherwise temperature control means such as thermoelectric coolers must be employed). Fortunately, distributed feedback, cleaved-coupled-cavity and other laser structures offer the narrow spectra required.

The multiplexing and demultiplexing of multiple wavelengths can most expeditiously be effected via passive devices. GRIN-rods (graded-index glass rods of the order of a mm in diameter) coupled to gratings (see Figure 7-1) offer the most ready implementation for large numbers of wavelengths. Compounds of dichroic mirrors suffice for less demanding numbers (see Figure 7-2).

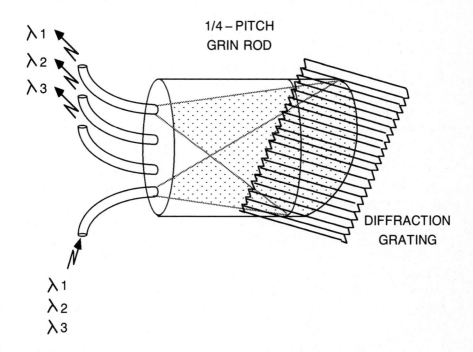

FIG. 7-1. GRIN-rod grating structure demultiplexing.

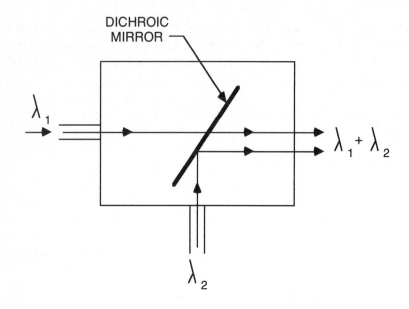

FIG. 7-2. Dichroic mirror structures.

Such multiplexing devices and circuits are only justifiable in applications where the trade-off with the cost of fiber is favorable. This becomes progressively more difficult as the losses associated with the multiplexing and demultiplexing increase, and as the costs of fiber decrease.

7.4 COHERENT DETECTION TECHNIQUES

Only a modest fraction of the bandwidth theoretically available to modulated light is utilized in amplitude-modulated applications; indeed, conventional modulation has been likened to intensity modulating a noise source. The carrier frequency of infrared light is of the order of 10^{14} Hz, and if the bandwidth it represents can be efficiently utilized, it will open up a new dimension in transmission. As a fortunate adjunct, the techniques that yield better bandwidth utilization also improve upon the receiver sensitivity.

There have been some successes in demonstrating *homodyning* and *hetero-dyning* [9], and the techniques are on the threshold of practical application (see Figure 7-3). Both techniques add a local oscillator signal to the incoming signal at the receiver, and recover the information content via filtering unwanted components after conversion to the electrical domain. In the case of homodyning, the local signal is at the same frequency as the carrier; for heterodyning, it is at a frequency significantly different from the carrier.

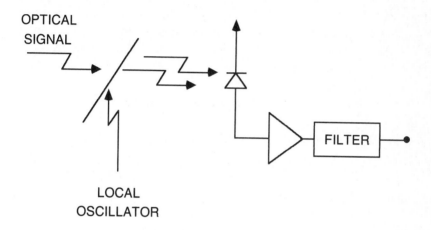

FIG. 7-3. Coherent detection techniques.

It can be shown [10] that homodyning could achieve the quantum limit discussed in Chapter 4 of a 10^{-9} BER with an average of 10.5 photons per bit if it were not for transmitting device phase noise. Heterodyning is 3 dB worse than homodyning. More specifically, taking into account the modulation technique, in comparison to Intensity Modulation with Direct Detection, Coherent Detection can provide a 3 dB improvement with FSK heterodyning, a 6 dB improvement with PSK heterodyning, and a 9 dB improvement with PSK homodyning [11].

Improvements in receiver sensitivity can be translated into increased range or decreased power demands upon the transmitter.

Difficulties still to be overcome in coherent techniques include: the need for ultra-stable light sources, inexpensive and effective modulators and demodulators, and polarization maintenance (it is important to efficient detection that the polarizations of the received and local oscillator signals match). Combining sources and modulating means in a single device in the manner that LEDs and ILDs are commonly employed appears unlikely to allow adequate stability;

external modulators are therefore likely necessary.

As an example, one experiment transmitted 400 Mbps signals over a 290 km link using an optimal FSK scheme called continuous-phase FSK [12].

For this application and for photonic switching (discussed in the next section), polarization-maintaining fiber is of importance. Fortunately, such fiber can be manufactured via, e.g., mechanical deformation that induces an elliptical cross-section that discourages one sense of polarization [13, 14]. Alternatively, local polarization-state control may be employed, utilizing means at the receiver to bring the local oscillator and received-signal polarizations together [15].

7.5 PHOTONIC SWITCHES

Switches capable of switching high frequency signals are difficult to fabricate using electronics alone. Crosstalk from other channels sharing a chip, circuit card, or frame can limit the minimum size of an electronic switch to dimensions that render it too expensive.

Switches have been made on exploratory bases which directly switch light. Such switches can enjoy almost complete immunity from adjacent-channel crosstalk, but may suffer from a measure of *on-off* crosstalk. Figure 7-4 displays a configuration employing an electro-optic material (lithium niobate) whose index of refraction can be altered by the application of an electric field. Light normally confined by one guide structure becomes free to couple into the adjacent structure when the requisite voltage is applied.

As a rule, the voltages necessary to operate such switching devices are lower if the light presented is polarized.

Ultimately, in most practical systems, the losses incurred in the light domain must be recouped somehow. Ideally, this amplification would take place in the light domain, reverting to the definition of LASER with its full implications (see Section 7.7). Some work has been done in this area, but the devices are likely to be at least as expensive as ILDs. For some time, therefore, excepting unusual applications, it may be more economical to bring the signal into the electrical domain for amplification and switching.

7.6 PHOTONIC LOGIC

Light domain switches and modulators can be employed to perform digital

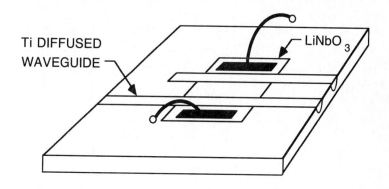

FIG. 7-4. A photonic switch.

logic functions as well, the attraction to such applications being the perception that higher speed switching should be possible, which would lead to higher performance digital systems.

It will be recalled that a *logically complete* set of elements is necessary in order to perform generalized logic functions such as are required in the fashioning of a digital computer. The operations AND, OR, and NOT (or INVERSION) form a complete set, as do either the NAND (NOT-AND) or NOR (NOT-OR) operations. (By way of contrast, the EXCLUSIVE-OR function does not constitute a complete set.)

The C^3 laser has proven capable of functioning as an OR, AND, INVERTER and EXCLUSIVE-OR [16].

Practical systems employing photonic logic to perform, for example, digital computation comparable to that available with electronic systems, must await comparably compact photonic logic devices. Such size reductions may prove difficult because photons are not as easily influenced as electrons. Immune to electrostatic and electromagnetic fields, photons require interaction with electrons in materials, and such interaction is so weak as to require devices spanning several wavelengths. Electronic logic devices with feature sizes less than the wavelength of infrared light are already becoming commonplace, and will be difficult to displace.

7.7 LIGHT-DOMAIN AMPLIFIERS

At the very least, it may be viewed as inelegant to be compelled over long transmission routes to convert from the optical domain to the electrical domain and back at repeaters because of the lack of an optical-domain amplifier. If possible and practical, remaining in the light domain might be best.

Electrical domain receivers contribute shot and dark-current noise from the detector and circuit noise from the amplifier, reducing the BER of the system. Further, they limit the system bandwidth and are subject to crosstalk from nearby systems.

Ideally, a photonic amplifier that would substitute for the detector, electrical-domain amplifier, driver and light source would be used. Though electrically powered, the repeating functions would all take place in the light domain.

Such amplifiers [17, 18] potentially exist in typical laser diodes. As has been described in an earlier chapter, below threshold current, an ILD behaves as an LED, producing spontaneous emission. Above threshold, it functions as a laser, producing stimulated emission. In the region between LED and ILD behaviors, the device is capable of functioning as a light amplifier capable of amplifying incident light signals from an external source.

Experimental amplifiers with gains of 30 dB and bandwidths exceeding 10 GHz have been reported [17], and it has been shown that a system with exclusively optical amplifiers could traverse 10,000 kilometers [19]. One experiment has successfully employed four light-domain (or optical) amplifiers distributed over 313 km with direct detection at 1 Gbps, and over 372 km using FSK and coherent detection [20]

Practical, economically feasible devices are not yet commercially available, but work is continuing.

7.8 INTEGRATED OPTICS

It is possible to accurately induce well-defined regions of higher refractive index via diffusing appropriate materials (e.g., titanium) into e.g., gallium arsenide or lithium niobate. Such regions can function as waveguides for conveying light between elements embedded in the medium. Such structures are referred to as *Monolithic Integrated Optics* (MIO) devices.

Among the elements comprising candidates for such structures are lasers (e.g., DBR types not requiring placement at the structure edge), detectors, associated circuitry, electro-optic devices such as couplers and combiners, etc.

Difficulties associated with building practical MIOs include the compromises sometimes necessary for differing devices to be fabricated on the same substrate, perfecting noncrosstalking crossovers, developing inexpensive external connections, yield problems, etc. A variation on this technique would be to mount devices upon the substrate, using it primarily for interconnection.

7.9 SOLITONS

Solitons are traveling pulses whose temporal shape does not change with time or distance. They were first noticed in association with water waves in a canal by J. Scott Russell in 1834 [21]. Russell studied the phenomenon for several years before presenting his findings before the Royal Society of Edinburgh.

These pulse types are of interest for transmission because they hold the promise of dispersionless communication: if a soliton pulse represents a binary "1," for example, it will retain that information over distance unsullied by effects other than attenuation.

A soliton's properties are explainable on the basis of a balance of two phenomena: nonlinear generation of harmonics, and dispersion. If the medium is nonlinear it will generate harmonics which would normally distort the pulse; a medium with complementary dispersive properties will eliminate these harmonics. The net effect is an undistorted pulse whose only enemy is attenuation.

Actually, solitons can also exist under conditions of linearity in a nondispersive medium [22].

In conventional glass fibers, nonlinear behavior is observed, but only at massive power levels [23]. Fibers displaying the requisite characteristics at moderate power levels are needed for practical application.

Solitons have been experimentally created via induced interaction between an ILD and reflected light from a fiber stub [24].

Soliton transmission may hold promise for future communications systems.

7.10 OPTICAL PHASE CONJUGATION

Some relatively recent results in the area of optical phase conjugation offer the possibility of further extending the bandwidth of fiber [25].

Phase-conjugate mirrors perform a "time-reversal"-like function in that they can cause light to reflect back to its source, taking the identical path it took on its way to incidence, effectively undoing the damage caused by aberrations in the path during the initial pass. Such a device, for example, might reproduce virtually exactly the same signal from the receiving-end of a fiber as was launched into it at the sending end if certain conditions were met.

Two techniques capable of producing phase conjugation are stimulated Brillouin scattering and four-wave mixing. Stimulated Brillouin scattering is induced by illuminating a material that has nonlinear characteristics at the power levels used such that a dynamic set of pressure density variations are created that have the net effect of reflecting the incident beam back upon itself as a conjugate beam. Four-wave mixing illuminates similar material with two reference beams and the object beam. The object beam interacts with one of the reference beams to produce an interference pattern from which the second reference beam is reflected to become the phase-conjugate beam.

A possible implementation of a fiber system utilizing such a device might consist of a span split at its midpoint with a phase-conjugate mirror employed to launch a phase-conjugated version of the signal at that point into the second half of the span. Note that the characteristics of the fiber employed for the second half would have to be identical to those of the first half.

The practicability and economy of using such a technique remain to be seen. The technique would likely be primarily applicable to very lengthy, high-bandwidth routes where the cost of additional equipment is compensated by the large channel cross-section.

7.11 IV-VI COMPOUNDS

The IV-VI and certain other compounds hold promise for longer wavelength light communication devices, matching the spectral characteristics of the ultra-low-loss halide glasses. In particular, the lead-salt materials (e.g., $Pb_{1-x}Sn_xTe$) are of interest [26]. Injection lasers utilizing these materials are capable of operation in the range of 3 to 34 μm.

7.12 NEW APPLICATIONS

New applications of fiber optics technology can at best be only speculated upon. Many eventual applications may be unimaginable today, others may simply be awaiting the advent of the low costs likely to come as manufacturing volume of potential constituents increases.

Among the fertile areas is sensing, an application in its infancy relative to fiber optics technology.

The application of Monolithic Integrated Opto-Electronics as described above to the integration upon a common substrate of optical sources, multiplexors and demultiplexors, switches, waveguides, detectors, and drive and receiver circuitry may also well become significant.

7.13 SUMMARY

Fiber optics systems are trending toward longer wavelengths, lower loss fiber, and higher bandwidth. Ultra-low-loss fiber is being studied, as is wavelength-division multiplexing (which has already attained modest utilization).

Coherent detection techniques hold the promise for longer spans in the immediate future, and much more efficient bandwidth usage in the more distant future.

Photonic devices capable of switching light and of performing logic are still in their infancy, with great promise for the future.

Light domain amplifier are beginning to reach commercial availability, and integrated optics are making strides.

Soliton transmission is still a curiosity, but may one day become practical.

IV-VI compound semiconductors may be developed to serve the spectral needs of the halide glasses.

EXERCISES

1. If the coupling and decoupling of each wavelength exacts a 3 dB penalty, how many wavelengths could reasonably be multiplexed (and demultiplexed) onto a fiber if the fiber were lossless?

2. Making reasonable assumptions about the losses involved in a light-domain switch, calculate the allowable loop lengths for a two-way video connection between two people served by the same video central office assuming no optical gain is provided.

3. Speculate upon the potential performance of a computer utilizing optical switching.

4. Show how a light-domain "AND," "OR," and "NOT" circuit might be built.

5. Suggest several parameters that would have potential for fiber sensing.

6. What light-launching device type would be most suitable for use in an integrated optics application?

7. Describe how a fiber-guided torpedo controlled from a submarine via a fiber paid out by the speeding weapon might be built. What problems might arise?

REFERENCES

[1] E. W. Mies and L. Soto, "Characterization of the Radiation Sensitivity of Single-Mode Optical Fibers," *11th European Conference on Optical Communication (ECOC)*, Venice, Italy, 1985.

[2] J. E. Midwinter, "Optical Fiber Communications, Present and Future," The Clifford Paterson Lecture, *Proceedings of the Royal Society of London*, A 392, 1984, pp. 247-77.

[3] F. P. Kapron, "Fiber-Optic System Tradeoffs," *IEEE Spectrum,* March, 1985, p. 70.

[4] S. E. Miller, "Present Thrust of Optical-Fiber Telecommunications Research," *IEEE Journal of Lightwave Technology: An Individual Perspective*, vol. LT-2, no. 4, August, 1984, pp. 494-95.

[5] L. G. Van Uitert, A. J. Bruce, W. H. Grodkiewicz, and D. L. Wood, "Minimum Loss Projections for Oxide and Halide Glasses." *Third International Symposium on Halide Glasses*, Rennes France, June, 1985.

[6] T. E. Bell, *IEEE Spectrum,* January, 1985, pp. 53-57.

[7] "Lightwave Multiplexer Combines Ten Laser Beams on Single Fiber," *AT&T Bell Laboratories Record,* March, 1985, p. 1.

[8] R. L. Bowmand and J. L. Lane, "Fiber Optics - An Exploding Industry," *Telephone Engineer and Management,* February, 1985, p. 115.

[9] R. Wyatt et al., "140 Mbit/s Optical FSK Fibre Heterodyne Experiment at 1.54 Microns," *Electronics Letters*, vol. 20, no. 22, October, 1984, p. 912-913.

[10]   J. Salz, "Coherent Lightwave Communication," *AT&T Technical Journal*, vol. 64, no. 10, December, 1985, pp. 2153-2210.

[11]   T. Okoshi, K Emura, K Kikuchi, and R. Th. Kersten, "Computation of Bit-Error Rate of Various Heterodyne and Coherent-Type Optical Communication Schemes," *Journal of Optical Communications*, vol. 2, no. 2, 1981, pp. 89-96.

[12]   K. Iwashita, et al., "Linewidth Requirement Evaluation and 290 km Transmission Experiment for Optical CPFSK Differential Detection," *Electronics Letters*, vol. 22, July, 1986, pp. 791-792.

[13]   S. C. Rashleigh and R. H. Stolen, "Status of Polarization-Preserving Fibers," *CLEO'84 Conference Proceedings*.

[14]   F. M. Sears and J. R. Simpson, "Polarization Quality of High-Birefringence Single-Mode Fibers," *AT&T Bell Laboratories Technical Journal*, February, 1984, vol. 63, no. 2, pp. 365-371.

[15]   T. Okoshi, "Recent Advances in Coherent Optical Fiber Communication Systems," *Journal of Lightwave Technology*, vol. LT-5, no. 1, January, 1987, pp. 44-52.

[16]   W. T. Tsang, N. A. Olsson and R. A. Logan, "Optoelectronic Logic Operations by C3 Lasers," *IEEE Journal on Quantum Electronics*, vol. QE-19, November, 1983, pp. 1621-5.

[17]   S. Kobayashi and T. Kimura, "Semiconductor Optical Amplifiers," *IEEE Spectrum*, May, 1984, pp. 26-33.

[18]   J. C. Simon, "Semiconductor Laser Amplifier for Single Mode Optical Fiber Communications," *Journal of Optical Communications*, April, 1983, pp. 51-62.

[19]   N. A. Olsson, "Single Frequency Lasers and 20 Gb/s Optical Transmission Experiments," *National Science Foundation Workshop and Grantee User's Meeting on Optical Communication Systems*, Ithica, New York, June, 1985.

[20]   N. A. Olsson et al., "Coherent and Direct Detection Transmission Experiments Using Optical Amplifiers," *OFC'88 Technical Digest*, New Orleans, Louisiana, January, 1988, p. THG4.

[21]   J. Scott-Russell, "Report on Waves," *Proceedings of the Royal Society of Edinburgh,* 1844, pp. 319-320.

[22]   G. E. Peterson, "Electrical Transmission Lines as Models for Soliton Propagation in Materials: Elementary Aspects of Video Solitons," *AT&T Bell Laboratories Technical Journal,* no. 6. Part 1, July-August, 1984, pp. 901-919.

[23]   R. G. Smith, "Optical Power Handling Capability of Low Loss Optical Fiber as Determined by Stimulated Raman and Brillouin Scattering," *Applied Optics,* vol. 11, November, 1972, pp. 2489-2494.

[24]   L. F. Mollenauer, "The Soliton Laser," *Physics Today,* January, 1985, pp. S-44-45.

[25]   B. Y. Zel'dovich, N. F. Pilipetsky, and V. V. Shkunov, *Principles of Phase Conjugation,* Springer-Verlag, 1985.

[26]   G. P. Agrawal and N. K. Duta, *Long-Wavelength Semiconductor Lasers,* Van Nostrand Reinhold, 1986, pp. 432ff.

APPENDIX A

APPARATUS AND INSTRUMENTATION

A.1 GENERAL

As in many fields and specialties, there exist a number of instruments and devices peculiar to fiber optics. Every well-equipped electronics laboratory facility will have any number of items of equipment generally useful for experimentation in that domain: power supplies, signal generators, work benches, oscilloscopes, voltmeters, etc. It should not be surprising, therefore, that experimentation associated with a new domain and technology may require some new or unusual equipment.

A.2 THE OPTICAL BENCH

For casual experimentation or design, an optical bench may be an unnecessary item. For serious experimenters, however, such a device may be a virtual necessity. Optical benches provide a stable surface and an almost arbitrary flexibility with respect to placement of fixtures.

PORTABLE MODELS

For light-weight activities, portable models exist which can be placed upon a conventional work bench. Such models do not contribute much to vibration elimination, but they have the typical array of apertures for convenient mounting of fixtures.

STATIC MODELS

Historically, optical benches were fashioned of granite because of its dimensional stability and great mass, making it relatively impervious to building vibrations. Similar benches are used in precision tooling machine shops for dimensioning checks.

Less expensive modern-day benches are typically steel and are floated on pneumatic cushions and foam rubber [1].

The choice of size and type of bench is a function of the requirements of expected experiments and budget constraints.

The weight of such benches can be so extreme as to pose a real structural danger to a building. Heavy-weight benches should not be installed without carefully surveying the structural capability of the laboratory floor and the route via which the bench will arrive (hallways, elevators, etc.).

A.3 LASERS

Though not a necessity, it is often convenient to have at least a Helium-Neon gas laser available; these devices are readily available from equipment suppliers [2]. The light produced is visible, and is of sufficient intensity and spot size so that it can launch considerable power into a fiber with indifferent alignment.

The wavelength of such lasers (633 nm) is nonideal for high-silica glass fibers, but it is close enough, and the beam intensity is sufficient such that it can be used with quite lengthy fibers.

A length of fiber (e.g., a km) is often used while coiled, and occasionally a hidden imperfection such as a crack greatly increases the net attenuation of the length. Such an imperfection is invisible when the fiber is illuminated by an infrared source but shows up as a bright spot when the HeNe source is used. Ohio Bell Telephone Company has begun using a HeNe laser to aid in finding fiber breaks in optical patch panels and in identifying a fiber among hundreds that may have become scrambled [3]. The glow at a break is sufficiently intense to show through protective tubing and colored jacketing, and a fiber bent around the index finger spills sufficient light for positive identification.

A line-up of optical devices can also be aligned visually with such a laser.

Often these lasers must be fired up at regular intervals in order to preserve their useful life.

WARNING:

HeNe lasers may be dangerous to retinal tissue. Great care should be exercised to avoid directly viewing the beam, and precautions should be taken to protect the unsuspecting visitor who may be unaware of the problem. A built-in key lock is often provided as a safety precaution; its purpose should not be defeated.

A.4 THE TIME-DOMAIN REFLECTOMETER

While not a typical piece of laboratory equipment, a Time-Domain Reflectometer can be very useful in locating fiber faults, poor splices and other mismatches [4, 5]. Further, because it generates pictorial information on the entire fiber (as opposed to a summary set of values), it can reveal more subtle characteristics, including instances of refractive index anomalies, large bubbles, etc.

The principle used is identical to that applied in metallic facility reflectometry. A narrow, high-intensity light pulse is launched into the fiber and the backscattered light is monitored. The attenuation of the fiber at the scanning wavelength can be readily determined, and the position of any discontinuity can be pinpointed.

Figure A-1 indicates a typical result. The traces are for two fibers of subtly differing characteristics which have been spliced together and scanned at two different wavelengths: 820 nm and 1300 nm. Note the attenuation in the individual fibers at the differing wavelengths: the attenuation differs as expected, but in addition, the second length of fiber, while having the same attenuation at 820 nm as the first length, has a lower attenuation at 1300 nm. Note also the splice- and end-reflections.

A massive reflection spike (not shown) occurs at the source-fiber interface, analogous to the powerful signal presented to a radar receiver when the outgoing pulse excites the dual-role (transmitting and receiving) antenna.

A.5 FLUX MEASURING METER

It is very useful to know quantitatively what the level of output (if any) a light source is producing. If a system does not work, it can be very difficult to track down the origin of the failure. A flux measuring meter can be quite useful in either circumstance.

A typical modern meter provides a direct digital readout in dBm. The sensing head must be selected to be appropriate to the wavelength of the light to be measured. Meter design may assume that a fiber will be available to insert into the pickup head; pigtailed sources are therefore convenient.

Measurements can be made at the source, at the end of the fiber span, or anywhere else desired.

A.6 ATTENUATORS

For dealing with arbitrary light levels, it is convenient to employ calibrated

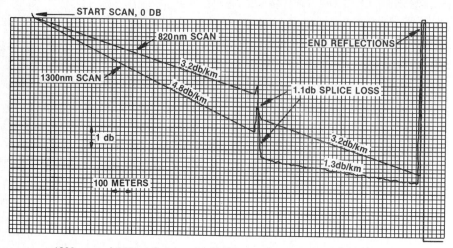

1300nm and 820nm Scans of a Spliced 1910 Meter Long Fiber Link

FIG. A-1. Time-domain reflectometer trace for two, spliced segments. (Reproduced with permission from ORIONICS, INC., *Application Note No. 8: OTDR ANALYSIS.*)

attenuators [6]. Similarly, such attenuators can serve (at least with respect to attenuation) as surrogates for lengths of transmission line (it can be cumbersome to have several kilometers of fiber spliced together for system test purposes).

Sometimes, as well, a system may in practice deliver to the receiver a signal that is so "hot" that the receiver is overloaded and distorts (e.g., a "standard" system being applied over an unusually short distance); an attenuator can rectify this problem.

AD HOC ATTENUATORS

As a quick and effective measure for reducing the intensity of a delivered signal, the fiber end may be roughened with sand or emery paper. (This technique was used, for example, in association with the 1977 lightwave trial in Chicago [7].)

Similarly, a connector may be loosened somewhat to introduce an airgap and provide some loss, but care is necessary to mechanically stabilize the loosened joint or vibration-sensitive attenuation may be experienced.

Another approach is to insert a small translucent but lossy piece of appropriate filter plastic into a connector.

FIXED ATTENUATORS

Attenuators are available on the market that provide a fixed, calibrated loss at a given wavelength. Some types are available in a conveniently cascadable form so that a combination of the attenuator blocks can provide a variety of net attenuations.

VARIABLE ATTENUATORS

There also exist continuously variable attenuators that permit selection of the required attenuation with the turn of a knob.

A.7 SPLICERS

For all but trivial spans, splices or connectors are a necessity. Often, as well, sources or detectors may have pigtail leads that must be coupled to a length of fiber.

AD HOC SPLICES

One of the more effective and convenient splicing mechanisms is the "loose-tube" technique [8]. This employs a segment of a square pipette tube with flared ends (see Figure A-2). Its dimensions are such that about four fibers could be snugly fit within (e.g., about 250-300 μm on a side).

The splicing operation consists of maneuvering two fibers (from either end) into the same corner of the tube about midway in its length and butting them together. Slight inward pressure is effective in preserving the butt.

This splice may be rendered permanent by applying, e.g., a drop of epoxy to either end.

Another technique is to effect temporary coupling between fiber ends via adjusting the position of one movable fiber end relative to another fixed fiber end. This function may be best performed with the aid of a microscope mechanical stage when one or the other of the fibers is illuminated by a HeNe laser source. Care should be exercised to prevent an excessive length of fiber from being cantilevered out from its anchoring point, or vibration may create a fluctuating coupling condition.

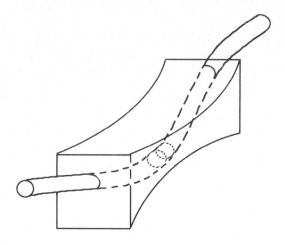

FIG. A-2. The loose-tube splice pipette.

FUSION SPLICERS

A permanent, low-loss splice may be obtained via fusing the two fiber ends together. Such a splice can be mechanically quite strong, and is the technique of choice for rugged conditions such as those to which undersea cable is subjected [9].

The most common heat source for this form of splicing is an electric arc, and arc-fusion splicers of several varieties are available on the market, often in portable form. A typical splicer of this type may have a preparation tool (e.g., a precision cutter) for readying the ends, a microscope for viewing the operation, micromanipulators for positioning, a means for coupling light into and out of the fiber to monitor position optimality, and a calibrated energy source for the arc [10].

There are some mechanical skills associated with the making of good splices even with the aid of such machines. Further, ambient conditions such as humidity levels can influence the losses associated with a splice (this is particularly important with respect to splicing operations in the field, which may take place in a manhole, on the ground, or on a telephone pole).

Highly automated portable fusion splicers utilizing microprocessor control are available at a premium price [11]. Characteristics of the fiber (single or multimode, dimensions, etc.) are keyed in and the operation becomes automatic.

As described in more detail in the text, adaptive butt splicers can provide remarkably good splices. Perhaps the best example is the AT&T actively-aligned rotary splice apparatus [12], which allows manipulation to guarantee the best possible splice.

A.8 INFRARED DETECTOR

Infrared detectors built on the principle of the "sniperscope" are a valuable laboratory adjunct. They employ a detector sensitive to infrared, whose output is used to electronically render a visible image on a tiny screen. Portable, hand-held models are available.

Such detectors may be employed, for example, to rapidly check whether an infrared source is functioning.

A.9 SAFETY EQUIPMENT

EYEWEAR

If conventional lasers or fiber optic systems operating at levels above approximately a milliwatt and using lens connections are to be employed, it is important to have safety eyewear available. Eyewear capable of attenuating specific ranges of wavelengths while providing reasonably good visibility are available [13].

WARNING SIGNS

Appropriate warning signs and labels of approved design (e.g., those specified in the ANSI Z136.1 [1986] standard) should be available and used.

A.10 FIELD CONNECTORS

Connectors have become available which can be applied to fiber ends in the field with relative ease. While generally somewhat inferior to connectors applied in a controlled production environment, they serve a useful purpose.

A.11 MISCELLANEOUS EQUIPMENT

SCRIBERS

With some practice, good fiber end preparation can be performed by simple breaking after scribing with a sharp tool. A silicon-carbide or diamond tip scribing tool is effective in this capacity. (Considerable practice may be necessary before good breaks can be routinely made.)

SOLVENTS AND HEATERS

Many fibers are jacketed with plastics which must be removed from the ends for, e.g., splicing operations. Mechanical removal holds the danger of damaging the fiber but may be effective if carefully done. Solvents may alternatively be used for stripping these jackets, e.g., dichloromethane; and for cleaning after stripping, e.g., ethanol or acetone.

MICROSCOPES

One or more moderate-power binocular microscopes can be used to good advantage in performing splices, preparing fiber ends, etc.

INDEX MATCHING FLUID

A small quantity of fluid appropriate for matching the index of refraction of the fiber being joined can reduce the loss of butt splices and connector joints, and increase tolerance to imperfectly prepared fiber ends. Siecor and General Electric are examples of manufacturers of this material.

MAGNETS

Though an apparently trivial footnote to useful equipment, a collection of small compliant permanent magnets of the kind often used for supporting papers upon metal walls can be very useful. The small size and light weight of uncabled fiber allow it to be firmly held against a metallic plate with such magnets for positioning activities.

EPOXY

Similarly, a ready supply of small-portion fast-setting epoxy can be very useful for rendering temporary splices permanent. If appropriately chosen, epoxies may serve as index-matching material.

MECHANICAL STAGES

Several mechanical stages mountable upon optical bench fixtures to allow fine multi-axis positioning can be quite useful if not essential. Linear translators, rotary tables, and angularly adjustable mirror and optical mounts which can be combined to give up to six degrees of freedom can be very useful. Motorized versions of such positioners are also available, allowing, for example, computerized position control.

FIXTURES

Fixtures such as hold-downs typical of optical-bench activities can prove very useful for supporting line-ups of optical componentry.

EXERCISES

1. Plan a laboratory for fiber-optics experimentation. Choose the equipment and, with the aid of a laboratory supply catalog, estimate the cost of such a facility; include the cost of floor-space.

2. Sketch the response that might be expected from a time-domain reflectometer scan of a fiber link comprised of two, 1 km lengths, the first of single-mode fiber with an attenuation of 1 dB/km, the second of graded-index fiber with an attenuation of 5 dB/km, butt-spliced together. (Both attenuations assumed to be at the wavelength of the time-domain reflectometer.)

3. What kind of background would you hope for in a laboratory technician for a fiber optics laboratory? Would Physics or Electronics training be of more value?

REFERENCES

[1]   E.g., ORIEL Corporation products (Stratford, Connecticut)

[2]   E.g., Spectra-Physics, Inc.

[3]   J. R. Aulicino, "Ohio Bell Lights Up FO Trouble Spots," *Telephony*, September 15, 1986, pp. 42-46.

[4] S. D. Personick, "Photon Probe - An Optical-Fiber Time-Domain Reflectometer," *Bell System Technical Journal,* vol. 56, no. 3, March, 1977, pp. 355-66.

[5] T. Horiguchi, "Optical Time Domain Reflectometer for Single-Mode Fibers," *Transactions of the Institute of Electronics and Communication Engineers, Japanese Sect. E.,* vol. 67, no. 9, September, 1984, pp. 509-515.

[6] E.g., Intelco Corp.

[7] H. Kressel, *Topics in Applied Physics*, vol. 39, Springer-Verlag, 1980, p. 266ff.

[8] C. M. Miller, "Loose Tube Splices for Optical Fibers," *Bell System Technical Journal*, vol. 54, no. 7, September, 1975, pp. 1215-1225.

[9] K. D. Fitchew, "Technology Requirements for Optical Fiber Submarine Systems," *IEEE Journal on Selected Areas in Communications,* vol. SAC-1, no. 3, April, 1983, p. 447.

[10] M. Hoshikawa, "Optical Fiber Splicing," *Japanese Annual Review of Electronic Computers and Telecommunications,* vol. 5, 1983, pp. 209-18.

[11] T. Haibara, "Fully Automatic Optical Fibre Arc-Fusion Splice Machine," *Electronics Letters,* vol. 20, no. 25-26, December, 6, 1984, pp. 1065-66.

[12] C. M. Miller and G. F. DeVeau, "Simple High-Performance Mechanical Splice for Single-Mode Fibers," *Technical Digest: Conference on Optical Fiber Communication*, San Diego, February 11, 1985, p. 21.

[13] "American National Standard for the Safe Use of Lasers," *American National Standards Institute Z136.1*, (ANSI, New York, 1986).

APPENDIX B

USEFUL
RELATIONS

B.1 GENERAL

Perusers of this text may come with any number of backgrounds. A typical student will have recently been exposed to electromagnetics and will feel comfortable with vector relations but may not be able to call them readily to mind. A practicing engineer has usually been away from such considerations for some time and will find "reminders" useful.

The following compendium is intended to summarize those relations most useful to the subject at hand. A number of other pertinent review topics are addressed as well in the interests of providing a self-contained treatment.

B.2 THE GRADIENT

The gradient provides a vector oriented in the direction of the greatest rate of change of a scalar field at a point.

$$\nabla A = \mathbf{i}\frac{\partial A}{\partial x} + \mathbf{j}\frac{\partial A}{\partial y} + \mathbf{k}\frac{\partial A}{\partial z}\ .$$

B.3 THE DIVERGENCE

The divergence of a vector quantity is a measure of the net flux to or from any closed surface about a point:

$$\nabla{\cdot}\mathbf{A} = \frac{\partial A_x}{\partial x} + \frac{\partial A_y}{\partial y} + \frac{\partial A_z}{\partial z}\ .$$

B.4 THE CURL

The curl is a somewhat more difficult function to explain. It is a measure of the circulation of a field, and is oriented along the axis of rotation.

$$\nabla \times \mathbf{A} = \begin{vmatrix} \mathbf{i} & \mathbf{j} & \mathbf{k} \\ \dfrac{\partial}{\partial x} & \dfrac{\partial}{\partial y} & \dfrac{\partial}{\partial z} \\ A_x & A_y & A_z \end{vmatrix}.$$

B.5 DEL SQUARED

Del squared is defined for a scalar argument, and for a vector argument in terms of the scalar argument expressions.

$$\nabla^2 = \frac{\partial^2}{\partial x^2} + \frac{\partial^2}{\partial y^2} + \frac{\partial^2}{\partial z^2}.$$

$$\nabla^2 \mathbf{A} = \mathbf{i}\nabla^2 A_x + \mathbf{j}\nabla^2 A_y + \mathbf{k}\nabla^2 A_z.$$

B.6 THE WAVE EQUATIONS IN CYLINDRICAL COORDINATE FORM

$$\frac{\partial^2 E_z}{\partial r^2} + \frac{1}{r}\frac{\partial E_z}{\partial r} + \frac{1}{r^2}\frac{E_z}{\partial \theta^2} = -\beta^2{}_T E_z$$

$$\frac{\partial^2 H_z}{\partial r^2} + \frac{1}{r}\frac{\partial H_z}{\partial r} + \frac{1}{r^2}\frac{H_z}{\partial \theta^2} = -\beta^2{}_T H_z$$

$$E_r = -\frac{j}{\beta_T{}^2}\left[\beta\frac{\partial E_z}{\partial r} + \omega\mu\frac{1}{r}\frac{\partial H_z}{\partial \theta}\right]$$

$$E_\theta = -\frac{j}{\beta_T{}^2}\left[\beta\frac{1}{r}\frac{\partial E_z}{\partial \theta} - \omega\mu\frac{\partial H_z}{\partial r}\right]$$

$$H_r = -\frac{j}{\beta_T{}^2}\left[\beta\frac{\partial H_z}{\partial r} - \omega\epsilon\frac{1}{r}\frac{\partial E_z}{\partial \theta}\right]$$

$$H_\theta = -\frac{j}{\beta_T{}^2}\left[\beta\frac{1}{r}\frac{\partial H_z}{\partial \theta} + \omega\epsilon\frac{\partial E_z}{\partial r}\right]$$

where, again,

$$B_T{}^2 = k^2 n^2 - \beta^2 .$$

B.7 POWER RATIOS: THE DECIBEL AND THE NEPER

A quantity expressed in dB *always* represents a power ratio. Though the powers themselves may be expressed in other terms (e.g., current, voltage), the dB nonetheless represents a ratio of powers:

$$\text{Power Ratio in dB} = 10 \log_{10} \frac{P_1}{P_0} ,$$

where P_1 is the power level of interest, and P_0 is the reference power level.

The dBm:

It is convenient to relate a power level to some reference power. One milliwatt is a common reference value, and a power level referenced to 1 milliwatt is indicated by expressing the ratio in dB as dBm.

A student schooled in modern digital design and familiar with the binary number system may find the following observation useful:

A power ratio of two corresponds to almost exactly 3 dB. Therefore a power ratio that is an integral power of 2 corresponds in dB to about 3 times that multiple.

EXAMPLE 1: A power ratio of 32, because it is equal to 2^5, corresponds to 15 dB.

EXAMPLE 2: 24 dB corresponds to a power ratio of 256 because $24/3 = 8$, and $2^8 = 256$.

It may be worthwhile in this context to remind that signal to noise ratios in the optical and electrical domains differ by a factor of two when expressed in dB (see Section 4.3).

The Neper:

A ratio can also be expressed in terms of the Napierian base:

$$\text{Ratio in nepers} = \ln\frac{A_1}{A_0},$$

where the subscripts again indicate value of interest and reference.

Note well, however, that the quantities in the ratio may or *may not* be powers, so that a ratio expressed in nepers must be examined carefully to determine what parameters the ratio pertains to.

B.8 BESSEL FUNCTIONS

Solutions for electromagnetic fields in structures with cylindrical symmetry inevitably involve Bessel functions (see Figure B-1).

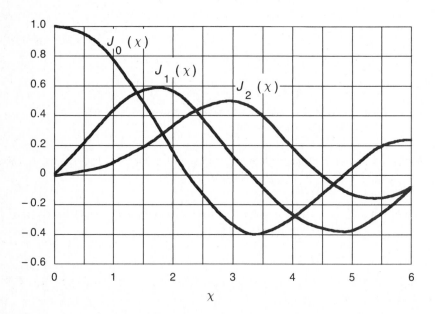

FIG. B-1. Bessel Functions of the first kind.

The differential equation encountered is of the form

$$\frac{d^2P}{d\rho^2} + \frac{1}{\rho}\frac{dP}{d\rho} + (1 - \frac{n^2}{\rho^2})P = 0 \ ,$$

where n is an integer (the more general case, where n is not restricted to integer values is of interest also).

Since the equation is of second order, two linearly independent solutions must exist. They can be found by assuming a power series form and substituting into the differential equation, generating series solutions.

The solutions are of two forms, the Bessel functions and the Neumann functions. The complete solution is then of the form

$$P = AJ_n(\rho) + BN_n(\rho) \ .$$

The two functions, $J_n(\rho)$ and $N_n(\rho)$, are referred to as Bessel functions of the first and second kind, respectively, of order n.

Note that since the Neumann functions become infinite at $\rho = 0$, they must be discarded as solutions for physical systems involving the origin (such as a fiber core).

Similarly, Bessel functions of the first kind are finite for infinite ρ, making them unsuitable for physical systems extending to infinity (such as the assumed extent of the cladding of an optical fiber in most analyses).

B.9 MODIFIED BESSEL FUNCTIONS

The modified Bessel equation of order n differs from the Bessel equation in that ρ is replaced by $j\rho$.

REFERENCES

There are many adequate references for this material, but the ultimate classic may be Schelkunoff's book:

S. A. Schelkunoff, *Electromagnetic Waves*, D. Van Nostrand Company, 1943.

APPENDIX C

SOLUTIONS TO EXERCISES

CHAPTER 1 EXERCISES

1-1 Power level diagram for 2, 1300 ohm loops.

26 Gauge cable has a resistance of 81.62 ohms/loop kft, and an attenuation of 2.67 dB/mile at 1 kHz.

$$\frac{1300}{81.62} = 15.93 \text{ loop kft.} \approx 3 \text{ miles}$$

Loss = $3 \times 2.67 = 8$ dB. See Figure C-1.

1-2 Coast-to-coast gain.

Coast-to-coast $\approx 3,000$ miles

16 Gauge loaded cable ≈ 0.14 dB/mile.

Gain needed $\approx 3,000 \times 0.14 = 420$ dB.

Cu. Weight = 7.818 lbs $\times 2$ per loop kft.

$3,000 \times 5.28 \times 7.818 \times 2 = 2.475 \times 10^5$ lbs ≈ 124 tons.

Cost (Cu only): $0.60/lb $\times 2,000$ lbs/ton $\times 124$ tons = $149k.

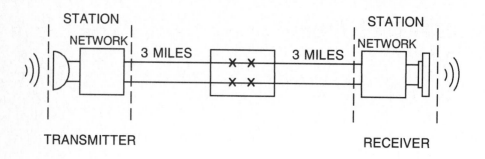

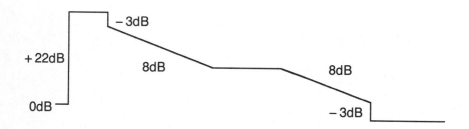

FIG. C-1. Figure for Exercise 1-1: Power levels for two 1300 ohm lines.

Glass Weight:

2.2 gms/cm^3 × π × (62.5 × 10^{-4} cm)^2× 3,000 $miles$ × 1.6×10^5 cm/$mile$

$\approx$ 1.3 × 10^2 kg $\approx$ 286 lbs.

1-3 Coast-to-coast repeaters.

Coast-to-coast $\approx$ 3,000 miles×1.6 km/mile = 4,800 km.

$$\frac{10^9 \; Hz-km}{4\times10^7 Hz} = 25 \; km.$$

$$\frac{4,800}{25} = 192 \text{ repeaters.}$$

Assuming repeater sees 1 μw in and provides 10 mW out, it provides a gain of 10^4 or 40 dB.

Loss $= 4,800 \times 0.25 = 1200$ dB.

$$\frac{1200}{40} = 30 \text{ repeater}$$

Though these are first-order calculations, they illustrate the relative significance of dispersion and attenuation limits for this case.

1-4 TV antenna rf photon flux.

$$\text{Power at set} = \frac{V^2}{R} = \frac{(10^{-3})^2}{300} \approx 3 \times 10^{-9} \text{ watts}.$$

$$\text{W/photon} = h\nu = \frac{3 \times 10^{-9} joules}{sec} \times sec = \frac{nh\nu}{sec} sec.$$

$$\frac{n}{sec} = \frac{P}{h\nu} = \frac{3 \times 10^{-9}}{6.63 \times 10^{-34} \times 6 \times 10^7} = \frac{10^{18}}{6.63 \times 2}$$

$$\approx 7.5 \times 10^{16} \text{ photons/sec.}$$

1-5 Antares carbon-dioxide laser.

12 trillion watts for 10^{-9} secs

$12 \times 10^{12} \times 10^{-9} = 12 \times 10^3$ joules.

$$\text{W/photon} = h\nu = h\frac{c}{\lambda} = 6.6 \times 10^{-34} \times \frac{3 \times 10^8}{10.6 \times 10^{-6}}$$

$$= 1.87 \times 10^{-20} \text{ joules/photon.}$$

Therefore, the number of photons $= \dfrac{12 \times 10^3}{1.87 \times 10^{-20}} = 6.4 \times 10^{23}$.

(Note the similarity to Avogadro's number.)

1-6 Sidewinder missile.

Assume: Electronic sensor in nose $\approx$ 10 gm
: Must raise temperature by $1000C^o$
: Heat capacity of the mass $= 0.5 \ kJ/kgC^o$

$$W_{needed} = \frac{0.5 \ kJ}{kg \ C^o} \times 10^3 C^o \times 10 \ gm = 5.0 \times 10^3 \text{ J.}$$

Assume: Illumination for 1 ms
: Diameter $= 3"$ (not the real dimension).

$$P = \frac{5 \times 10^3 J}{10^{-3} sec} = 5 \times 10^6 \text{ watts.}$$

$$\text{Area} = \pi \times (2.5 \ in \times 0.0254 \ m/in)^2 = 1.27 \times 10^{-2} m^2.$$

$$P/m^2 = \frac{5 \times 10^6}{1.27 \times 10^{-2}} = 3.93 \times 10^8 \ watts/m^2.$$

$$P/m^2 = EH = \frac{E^2}{377} \quad \text{(Using } E = 377 \text{ H)}$$

$$E = \sqrt{377 \times 3.93 \times 10^8} = 3.85 \times 10^5 \text{ volts/m}$$

$$H = \frac{3.85 \times 10^5}{377} \approx 10^3 \text{ amps/m.}$$

Air Breakdown at sea-level $\approx 5 \times 10^5$ volts/m.

Note that the above calculations were on the basis of simple calorimetry. In actual practice, a powerful laser beam will induce hypersonic shock waves in irradiated materials, causing ablation out of proportion to calorimetric considerations.

1-7 Circular waveguide vs. fiber.

BW of circular waveguide $\approx$ 70 GHz

Cost $\approx$ \$10/m

Specific Cost $\approx \dfrac{\$10}{70 \times 10^9} = 1.43 \times 10^{-10} \dfrac{\$}{Hz \cdot m}$

BW of fiber $\approx$ 2 GHz

Cost $\approx$ \$0.30/m

Specific Cost $\approx \dfrac{0.30}{2 \times 10^9} = 1.5 \times 10^{-10} \dfrac{\$}{Hz \cdot m}$

Observe that (for the assumptions made), the two numbers are very similar.

Note: True comparison requires consideration of the costs and spacing of repeaters.

1-8 Geostationary satellite orbit.

$$F = \frac{GMm}{r^2} = \frac{mv^2}{r}$$

$$r = \frac{GM}{v^2}$$

$$G = 6.67 \times 10^{-11} \frac{Nm^2}{kg^2}$$

$$M = 6 \times 10^{24} \text{ kg}$$

$$v = \frac{2\pi r}{24 \times 3600}$$

$$r = \frac{GM}{\left[\dfrac{2\pi r}{24 \times 3600}\right]^2}$$

$$r^3 = 76 \times 10^{21}$$

$$r = 4.25 \times 10^7 m = \frac{4.25 \times 10^7}{1.6 \times 10^3} = 2.66 \times 10^4 \text{ miles}$$

Subtracting the 4,000 mile radius of the earth yields 22,600 miles.

$$\text{Linear velocity} = \frac{2\pi r}{24} = \frac{2\pi \times 26,600}{24} = 6964 \text{ mph } (= 11,142 \text{ km/h}).$$

1-9 Constants.

$$\$_m = \frac{k_3}{L^2}$$

Using the results of Exercise 1.7,

$$1.43 \times 10^{-10} \frac{\$}{Hz \cdot m} = \frac{k_3}{(10^{-3} \frac{dB}{m})^2}$$

$$k_3 = 1.43 \times 10^{-16} \frac{\$ \ (dB)^2}{m^3 \ Hz}$$

$$\$_f = \frac{k_4}{B}$$

$$1.5 \times 10^{-10} \frac{\$}{Hz \cdot m} = \frac{k_4}{2 \times 10^9 \ Hz}$$

$$k_4 = 0.3 \frac{\$}{m} \ .$$

CHAPTER 2 EXERCISES

2-1 Snell's law.

The phrase "basic considerations" can be interpreted several ways, and the law can in fact be proven using Maxwell's Equations. What will be shown is a geometric optics analysis. See Figure C-2.

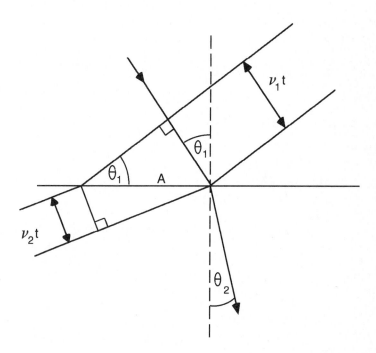

FIG. C-2. Figure for Exercise 2-1.

$A \sin \theta_2 = v_2 t$

$A \sin \theta_1 = v_1 t$

$$A = \frac{v_2 t}{\sin \theta_2} = \frac{v_1 t}{\sin \theta_1}$$

$$\frac{\sin \theta_1}{\sin \theta_2} = \frac{v_1}{v_2}$$

$$n_1 = \frac{c}{v_1}; n_2 = \frac{c}{v_2}$$

$$\frac{\sin \theta_1}{\sin \theta_2} = \frac{c/n_1}{c/n_2} = \frac{n_2}{n_1}$$

$$n_1 \sin \theta_1 = n_2 \sin \theta_2.$$

2-2 Fiber cost estimate.

Assume:

> Cost/Price = 1/2
> Overhead = 1/2
> Cost of Cabling = Cost of Fiber
> Materials = Labor

Cost = $0.05/m

Cabling = Fiber = $0.025/m

10 km/preform = $2.5 \times 10^{-2} \times 10^4$ = $250

Overhead = $125

Materials = Labor = $62.50

Labor for 10 hours = $6.25/hour.

2-3 Preform diameter calculation.

See Figure C-3.

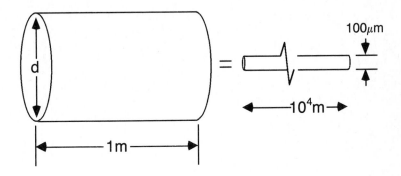

FIG. C-3. Figure for Exercise 2-3.

$\pi(d/2)^2\times1 = \pi(50\times10^{-6})^2\times10^4$

$d^2/4 = 2500\times10^{-12}\times10^4$

$d^2 = 10^4\times10^{-8} = 10^{-4}$

$d = 10^{-2}m = 1 \text{ cm.}$

2-4 Embedded lightsource.

See Figure C-4.

THE HARD WAY:

$n_1\sin \theta_1 = n_2\sin \theta_2$

$\sin \theta_c = n_2/n_1$

$\sin \theta_s = \cos \theta_c = \sqrt{1-\sin^2 \theta_c}$

$\qquad = \sqrt{1-(n_2/n_1)^2}$

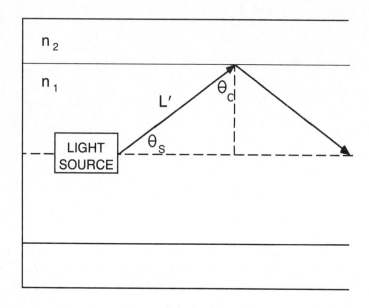

FIG. C-4. Figure for Exercise 2-4.

$$= \sqrt{\frac{n_1^2 - n_2^2}{n_1^2}}$$

$$= \sqrt{2\Delta}$$

$$= \sqrt{\frac{(NA)^2}{n_1^2}} = \frac{NA}{n_1}.$$

Note: Smaller than NA, why? (Because the focusing effect of the front-face refraction is lost.)

THE EASY WAY

Use Eqn. (2.9)

$$\sin \theta_0 = n_1/n_0\sqrt{1-[(n_2/n_1)\sin \theta_2]^2}$$

Note that $n_1 = n_0$, so

$$\sin \theta_0 = \sqrt{1-[(n_2/n_1)\sin \theta_2]^2}$$

and $\theta_2 = \pi/2$, so

$$\sin \theta_s = \sqrt{1-(n_2/n_1)^2}.$$

2-5 Intrinsic region transit time.

The highest possible NA is sin $(\pi/2) =1$. For a core index (n_1) of $\sqrt{2}$, the cladding index would have to be 1 to produce an NA of 1. Only free space, or, as a good approximation, air, would serve this purpose. This implies a fiber consisting only of a core suspended in air. Such a configuration would only be practical in a very short run in a protected environment needing no supporting structure. The bandwidth would theoretically approach zero because rays virtually orthogonal to the axis would be accepted, but would probably be of usable magnitude for most practical sources.

2-6

At the dispersion zero, the first term in the denominator of Equation 2.2 becomes zero, so that:

$$B = \frac{1}{8S(\lambda)L(\Delta\lambda)^2}$$

From figure 2-18, the slope is about 32 ps for 0.3 μm. Substituting,

$$B \approx \frac{1}{8\times\dfrac{32ps/nm-km}{0.3\times10^3 nm}\times1km\times(2nm)^2}$$

$$\approx \frac{0.3\times10^3}{8\times32\times4\times10^{-12}}.$$

Noting that $8 \times 32 \times 4 = 2^{10} \approx 10^3$,

$B \approx 0.3 \times 10^{12} = 300$ GHz.

2-7 Modal dispersion delay dependence on NA.

See Figure C-5.

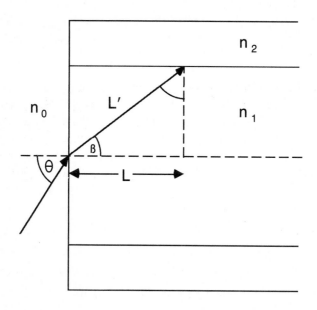

FIG. C-5. Figure for Exercise 2-6.

$$\frac{L'}{L} = \frac{1}{\cos \beta}$$

$$n_0 \sin \theta = n_1 \sin \beta$$

$$\cos \beta = \sqrt{1 - \sin^2 \beta}$$

$$\frac{L'}{L} = \frac{1}{\sqrt{1 - \sin^2 \beta}} = \frac{1}{\sqrt{1 - (n_0/n_1)^2 \sin^2 \theta}}$$

$$n_0 = 1; \quad velocity = \frac{c}{n_1}$$

For $\theta =$ the acceptance angle,

$$\frac{L'}{L} = \frac{1}{\sqrt{1 - \frac{(NA)^2}{n_1^{\,2}}}}$$

$$\approx \frac{1}{1 - \frac{1}{2}\frac{(NA)^2}{n_1^{\,2}}} \;,$$

using $\sqrt{1 - x} \approx 1 - \frac{x}{2}$ for x small.

$$\delta = \tau' - \tau$$

$$= \frac{L'n_1}{c} - \frac{Ln_1}{c} \;,$$

using $\tau = \dfrac{L}{velocity} = \dfrac{Ln_1}{c}$

$$\delta = \frac{n_1}{c}(L' - L)$$

$$= \frac{n_1 L}{c}\left(\frac{L'}{L} - 1\right)$$

$$\approx \frac{n_1 L}{c}\left[\frac{1}{1 - \frac{1}{2}\frac{(NA)^2}{n_1^{\,2}}} - 1\right]$$

$$\approx \frac{n_1 L}{c}\left[\frac{1 - 1 + \frac{1}{2}\frac{(NA)^2}{n_1^{\,2}}}{1 - \frac{1}{2}\frac{(NA)^2}{n_1^{\,2}}}\right]$$

$$\approx \frac{n_1 L}{2c} \times \frac{(NA)^2}{n_1^2}$$

and for $(NA)^2 \ll 2n_1^2$,

$$\approx \frac{L}{2c} \times \frac{(NA)^2}{n_1}.$$

Another Approach

Attacking the problem from another point, note that when α is the critical angle,

$$\frac{L}{L'} = \sin \alpha = \frac{n_2}{n_1}$$

Noting that $\dfrac{(NA)^2}{n_2} = \dfrac{(n_1 + n_2)(n_1 - n_2)}{n_2}$

$$\approx \frac{2n_2(n_1 - n_2)}{n_2} = \frac{n_1 - n_2}{2}$$

for $n_1 \approx n_2$, the same result is obtained.

2-8 Fiber core diameter for single-mode operation.

Given: $\lambda = 1 \ \mu m$, $n_1 = 1.4$, and $\Delta = 0.0014$.

For single-mode operation, $a < \dfrac{1.2\lambda}{\pi(n_1^2 - n_2^2)^{1/2}}$

$$\Delta = \frac{n_1^2 - n_2^2}{2n_1^2}$$

Substituting, $a < \dfrac{1.2\lambda}{\pi n_1 (2\Delta)^{1/2}} \approx 5.16 \times 10^{-6}$,

Therefore, the diameter must be less than ≈ 10.32 μm.

2-9 Thin cladding materials savings.

$$V = ka\,(n_1{}^2 - n_2{}^2)^{\frac{1}{2}}$$

$$k = \frac{2\pi}{\lambda}$$

Substituting the critical value for V,

$$2.4 = \frac{2\pi}{\lambda}a\,\sqrt{n_1{}^2 - n_2{}^2}$$

$$\Delta = \frac{n_1{}^2 - n_2{}^2}{2n_1{}^2}$$

$$\sqrt{n_1{}^2 - n_2{}^2} = \sqrt{2n_1{}^2\Delta}$$

$$= n_1\sqrt{2\Delta}$$

$$a = \frac{1.2\times10^{-6}}{\pi}\times\frac{1}{1.4\sqrt{0.028}}$$

$$= 1.6\times10^{-6}\ (\text{or diameter} = 3.2\ \mu\text{m}).$$

$$\frac{b}{a} = \frac{1}{2W}\,\ln\left\{\frac{8.7W\lambda[1-(W/V)^2]}{2\pi a^2 n}\right\} - \frac{1}{2W}\ln\alpha$$

$$W = \frac{V}{\sqrt{2}},\ a = \frac{V}{k\,(NA)},\ V = 2.4,\ \text{and}\ k = \frac{2\pi}{\lambda}$$

$$\frac{b}{a} = \frac{1}{\sqrt{2}V}\ln\left\{\frac{8.7\lambda4\pi^2(NA)^2}{4\sqrt{2}\pi V n\lambda^2}\right\} - \frac{1}{\sqrt{2}V}\ln\alpha$$

$$= \frac{1}{\sqrt{2}V}\ln\left\{\frac{8.7\pi(NA)^2}{\sqrt{2}V n\lambda}\right\} - \frac{1}{\sqrt{2}V}\ln\alpha$$

$V = 2.4$, $NA = 0.1$, $n = 1.4$, $\lambda = 10^{-6}$

$$\frac{b}{a} = \frac{1}{\sqrt{2}\times2.4}\ln\left\{\frac{8.7\pi\times.01}{\sqrt{2}\times2.4\times1.4\times10^{-6}}\right\} - \frac{1}{\sqrt{2}\times2.4}\ln\alpha$$

$$= 3.18 - 0.29 \ln \alpha$$

For $\alpha = 0.25$, $\dfrac{b}{a} = 3.58$

For $\alpha = 5.0$, $\dfrac{b}{a} = 2.71$

The volume is proportional to r^2:

$(3.58)^2 = 12.82$

$(2.71)^2 = 7.34$

The savings are therefore $\dfrac{5.48}{12.82}$, or 43%.

2-10 Ultra-low attenuation fiber.

1,000 mile transmission

a. Assume

Launched power = 0 dBm.

Receiver sensitivity = -30 dBm

$$\frac{30}{1600 \ km} = 0.02 \ dB/km.$$

b. How low must splice loss be to be negligible?

Say splice every 10 km: 160 splices

Say 10 dB for splices: $\dfrac{10}{160}$ = 0.06 dB/splice.

c. Cu pair α to be -70 dB at $\approx$ 2,000 miles

Assuming 0 dBm in,

$\dfrac{70}{3200}$ km = 0.02 dB/km.

Note the similarity to (a) above.

2-11 Core-cladding refractive index discontinuity.

Check Eqn. 2.12 for discontinuity at boundary.

$$n(r) = n_1 \left[1 - 2\Delta \left(\frac{r}{a} \right)^{\alpha} \right]^{\frac{1}{2}}$$

$$\Delta = \frac{n_1{}^2 - n_2{}^2}{2n_1{}^2}$$

$$n(0) = n_1 \left[1 - 2\Delta(0)^{\alpha} \right]^{\frac{1}{2}} = n_1$$

$$n(a) = n_1 \left[1 - 2 \times \left[\frac{n_1{}^2 - n_2{}^2}{2n_1{}^2} \right] \times (1)^{\alpha} \right]^{\frac{1}{2}}$$

$$= n_1 \left[\frac{n_1{}^2 - n_1{}^2 + n_2{}^2}{n_1{}^2} \right]^{\frac{1}{2}} = n_2.$$

What would be the effects of discontinuities?

See Figure C-6.

As can be seen, there are two extremes. In case a, the abrupt change in refractive index would approximate the depressed cladding structure often used with

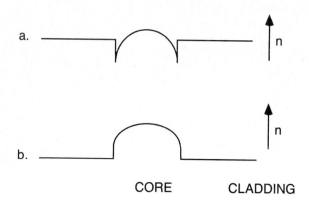

FIG. C-6. Figure for Exercise 2-9.

single-mode fiber. In case b, the boundary would approach the characteristics of a step-index fiber and would reduce the bandwidth because a few errant sluggish modes would tend to be supported.

2-12 Low-order slab modes.

One way in which to get an approximate solution is to make inspired guesses; after a little manipulation of Equation 2.16 such a guess will be tried.

let $g = \dfrac{2\pi h}{\lambda}$, and $x = n_1 \cos \theta$, then

$$gx^2 - 2\tan^{-1}\sqrt{n_1{}^2 - n_2{}^2 - x^2} = \nu\pi x.$$

Now we would expect θ to be very close to $\pi/2$ for the lowest order modes, so that $\cos^2 \theta$ should be very small indeed. We will assume then that the x^2 term under the radical is small compared to the difference between the squares of the indices of refraction. (This assumption can be tested afterward to see if it is consistent with the result.)

Subject to this assumption,

$$gx^2 - v\pi x - 2\tan^{-1}(NA) = 0.$$

Solving for x,

$$x = \frac{v\pi \pm \sqrt{(v\pi)^2 + 8gtan^{-1}(NA)}}{2g}.$$

Solving for $\cos\theta$,

$$\cos\theta = \frac{v\pi + \sqrt{(v\pi)^2 + 16\pi(h/\lambda)\tan^{-1}(NA)}}{4\pi hn_1/\lambda}.$$

Now, since we know that a simple ray analysis only applies for wavelengths small compared to the dimensions of the guide, let us assume that h/λ is 10, that NA = 0.23, and that $n_1 = 1.4$.

Calculating, we find that for $v = 0$, $\theta = 86.5$ degrees, and for $v = 1$, $\theta = 85.4$ degrees. Checking on our initial simplifying assumption, we conclude that it was justified.

2-13 Step-index fiber modes.

About how many modes will a 50 μm core, step-index fiber with Exercise 8 properties support when illuminated by a 1 μm source?

$n_1 = 1.4$ and $\Delta = 0.0014$

$$\text{Modes} \approx \frac{V^2}{2}$$

$$V = ka\sqrt{n_1^2 - n_2^2}$$

$$\Delta = \frac{n_1^2 - n_2^2}{2n_1^2}$$

Therefore $V = kan_1\sqrt{2\Delta}$

$$\text{Modes} \approx \frac{\left[\frac{2\pi}{\lambda}an_1\sqrt{2\Delta}\right]^2}{2}$$

$$\approx \left[\frac{2\pi}{\lambda} a n_1 \right]^2 \times 0.0014$$

$$\approx (70\pi)^2 \times 0.0014$$

$$\approx 68 \text{ Modes.}$$

2-14 Equation 2.44 verification.

Does $V^2 = U^2 + W^2$?

$$U = \sqrt{n_1^2 k^2 - \beta^2} \; a$$

$$W = \sqrt{\beta^2 - n_2^2 k^2} \; a$$

$$V^2 = (n_1^2 k^2 - \beta^2 + \beta^2 - n_2^2 k^2) a^2$$

$$= k^2 a^2 (n_1^2 - n_2^2) = [ka(NA)]^2 .$$

Yes.

CHAPTER 3 EXERCISES

3-1 ILD gain.

See Figure C-7.

$$g = \alpha + \frac{1}{2L} \ln \frac{1}{R_1 R_2} \quad \text{(in nepers)}$$

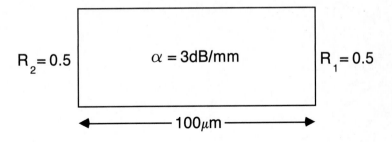

$R_2 = 0.5$ $\alpha = 3\text{dB/mm}$ $R_1 = 0.5$

$\longleftarrow$ 100μm $\longrightarrow$

FIG. C-7. Figure for Exercise 3-1.

$10 \log_{10} \dfrac{A}{A_0}$: dB

$\ln \dfrac{A}{A_0}$: nepers

3 dB/mm = 3000 dB/m = 693 nepers/m

$$g = 693 + \frac{1}{2 \times 10^{-4}} \ln \frac{1}{\frac{1}{2} \times \frac{1}{2}}$$

$\quad = 693 + 6931 = 7624$ nepers/m

$\quad = 0.7624$ nepers over 100 μm

$\quad =$ a factor of 2.14.

(The actual GaAs - Air interface reflectivity is about 0.32.)

3-2 Internal ILD critical angle.

$$\sin \theta_c = \frac{3.4}{3.6} = 0.94$$

$\theta_c = 71^0$.

3-3 P-i-n quantum efficiency.

$$\eta = \frac{I_p/q}{P_{opt.}/h\nu}$$

$$= \frac{0.5\times10^{-6}h\nu}{10^{-6}q} = \frac{0.5h\nu}{q} = \frac{0.5hc}{q\lambda}$$

$$\eta = \frac{0.5\times6.6\times10^{-34}\times3\times10^8}{1.6\times10^{-19}\times10^{-6}}$$

$$= \frac{3.3\times3\times10^{-26}}{1.6\times10^{-25}}$$

$$= \frac{9.9\times10^{-1}}{1.6} = 0.62$$

$$R = \frac{I_p}{P_{opt}} = \frac{0.5\times10^{-6}}{10^{-6}} = 0.50 \text{ amps/watt.}$$

3-4 Infinite APD decay time.

See Figure C-8. The diagram illustrates that if the ionization coefficients of the two particle species are similar, continual avalanching, once triggered by an impinging photon, might continue ad infinitum unless the power supply is removed. In practice, the closer the coefficients are to the same value, the longer the recovery time when light ceases to impinge, and the lower the effective bandwidth.

The differing slopes correspond to the differing mobilities. (Such considerations are discussed, e.g., in J. Gowar, *Optical Communication Systems*, Prentice-Hall, 1984, p. 393.)

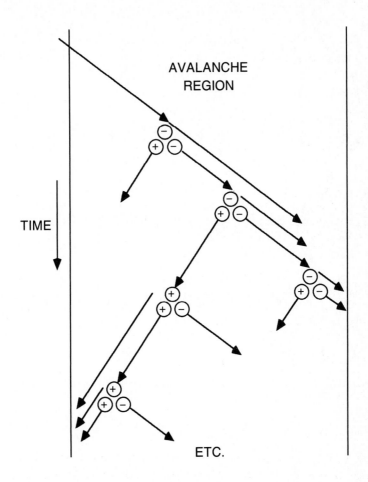

FIG. C-8. Figure for Exercise 3-4.

3-5 P-i-n intrinsic region transit time.

$\mu = 10^3 \ cm^2/Vs$

Velocity $= \mu E$

$E = V/d = \dfrac{5}{5 \times 10^{-6}} = 10^6 \ V/m = 10^4 \ V/cm$

Velocity $= 10^3 \times 10^4 = 10^7$ cm/sec (In the vicinity of saturation velocity.)

$$t = d/vel = \frac{5 \times 10^{-4}}{10^7} = 5 \times 10^{-11} \text{ sec}$$

$$= 0.05 \text{ ns}$$

More directly:

$$t = \frac{d}{vel} = \frac{d}{\mu E} = \frac{d^2}{\mu V}$$

$$= \frac{(5 \times 10^{-4})^2}{10^3 \times 5} = \frac{25 \times 10^{-8}}{5 \times 10^3} = 5 \times 10^{-11} \text{ sec} .$$

3-6 Raman-scattering power threshold.

$$P = 10^8 \times \pi r^2 \quad (\text{r in cm})$$

$$= \pi (5 \times 10^{-4})^2 \times 10^8$$

$$= 25\pi \approx 78 \text{ watts}$$

Expressed in dBm: $10 \log \dfrac{78}{10^{-3}} = 48.9$ dBm.

Note that this could also be calculated using the technique of Appendix B: $78,000 \approx 2^{16}$, and $3 \times 16 = 48$ dBm.

Assuming 0.2 dB/km fiber, a receiver sensitivity of -40 dBm, and neglecting other losses, the range would be

$$\frac{88}{0.2} = 440 \text{ km.}$$

3-7 Source coupling efficiency.

$$\text{Eff.} = \frac{P_{fiber}}{P_{source}}$$

$$= \frac{\displaystyle\int_0^{\sin^{-1}NA} (cos\ \theta)^n \sin\theta\ d\theta}{\displaystyle\int_0^{\pi/2} (cos\ \theta)^n \sin\theta\ d\theta}$$

$$= \frac{\dfrac{-1}{n+1} \cos^{n+1}(\theta)\ \Big|_0^{\sin^{-1}NA}}{\dfrac{-1}{n+1} \cos^{n+1}(\theta)\ \Big|_0^{\pi/2}}$$

$$= \frac{[cos\ (\theta_{NA})]^{n+1} - 1}{0 - 1}$$

$$= 1 - \cos^{n+1}(\theta_{NA})$$

$$= 1 - \left[\sqrt{1 - \sin^2(\theta_{NA})}\right]^{n+1}$$

$$= 1 - \left[\sqrt{1 - (NA)^2}\right]^{n+1}$$

$$= 1 - [1 - (NA)^2]^{\frac{n+1}{2}}$$

Check for n = 1:

$$= (NA)^2\ ;$$

and for n = ∞:

$$= 1\ .$$

3-8 Dual function, source-detector device.

The LED must employ III-V materials in a p-n junction. Silicon and Germanium, while attractive for detectors, are poor for light launching because they are indirect bandgap materials.

It would be constructed as a surface emitter in order to have a reasonable quantum efficiency when used as a detector.

The device would be less efficient and slower as a detector because there would be no intrinsic region separating the p and n materials. A low quantum efficiency, high leakage current, and high capacitance could be expected.

It might be usable in less demanding roles, and might be cheaper because of the higher volume (×2) and the advantages of having only one part to stock.

The proposal to fabricate such a device is not pure whimsy. A discussion of a system utilizing this property can be found in R. I. MacDonald, "Bidirectional Analogue Optical Transmission Using Semiconductor Junction Transceivers," *Electronics Letters*, vol. 15, no. 4, February, 1979, pp. 121-123.

3-9 Helium-Neon laser precautions.

Precautions for HeNe laser:

> Keep custody of the key to the laser if it has one.

> Place warning signs with the appropriate message on walls and outside of door as per ANSI Z136.1 (1986).

> Cover glass in door if present.

> Use the laser in a shielded space in the laboratory.

> Employ protective eyewear when appropriate.

CHAPTER 4 EXERCISES

4-1 Common-base LED drive.

See Figure C-9. Common-base configurations are capable of voltage gain but not current gain, and net power gain. Their major advantages would be input-to-output isolation without Miller Effect limitations, freedom from the potential oscillation problems of the emitter follower, and the ability to respond to low input voltage swings.

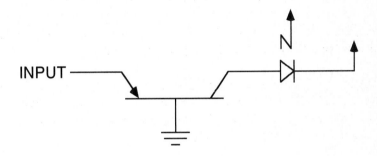

FIG. C-9. Common-base LED drive.

4-2 ILD Threshold

Below threshold, $I \approx I_0[e^{k(V - IR)} - 1]$.

Above threshold, $V \approx E_g/q + IR$.

Substituting, $V \approx E_g/q + I_0R[e^{kE_g/q} - 1]$.

To a good approximation, $\lambda = 1.2/E_g$, with λ in microns and E_g in electron volts. Then, for vanishing R,

$$V \approx \frac{1.2}{\lambda}.$$

For $\lambda = 1.3$ μm, $V = 0.92$ volts, and for $\lambda = .82$ μm, $V = 1.46$ volts. The IR drop at threshold will typically amount to of the order of an additional volt.

4-3 Linearization

See Figure C-10. For well-matched LEDs, the result can be expected to be identical to that of a push-pull amplifier: elimination of second-harmonic distortion components. Again, with properly matched components, the same technique should work at two different wavelengths over the same fiber.

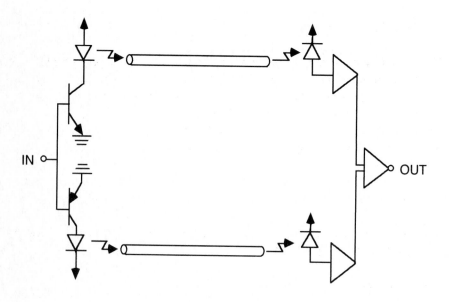

FIG. C-10. A linearization scheme.

4-4 Feedforward

See Figure C-11. What is shown is an attempt at a hybrid light-domain feedforward amplifier, where a portion of the signal processing is performed in the electrical domain. The input signal modulates the light-emitting diode, and the resulting light output is tapped, converted back to an electrical signal and compared with the original input signal. The difference is inverted and used to drive another, presumably matched diode whose output is combined additively with the original, corrupted optical signal.

The argument is that the corruption signal, being small, will introduce a lower order corruption when it is converted to optical form. The second light-emitting diode may therefore not have to be matched with the first one so long as its characteristics are not extraordinarily nonlinear. (It would, however, have to be fairly closely matched with respect to output wavelength.)

4-5 New Drive Circuit

There are any number of possible drive circuits that might be "invented." One candidate circuit is shown in Figure C-12. A transformer has the advantage of being able to provide a high-current output with low-current input, but some means must be provided to allow recovery of the state of the magnetic material (if used). A non-emitting diode is provided for that purpose. The turns ratio could, in some applications, be juggled to advantage.

4-6 Incident Power Level

Assuming that ones and zeroes are equally likely, 1,000 photons per one corresponds to 500 photons per bit.

$E = h\nu = hc/\lambda$, therefore

$$P = \frac{1,000 \text{ photons} \times 6.6 \times 10^{-34} J/Hz \times 3 \times 10^8 m/s \times 100 \times 10^6 bps}{2 \times 10^{-6} m}$$

$\approx 10^{-8}$ watts or -50 dBm.

4-7 Transimpedance Amplifier

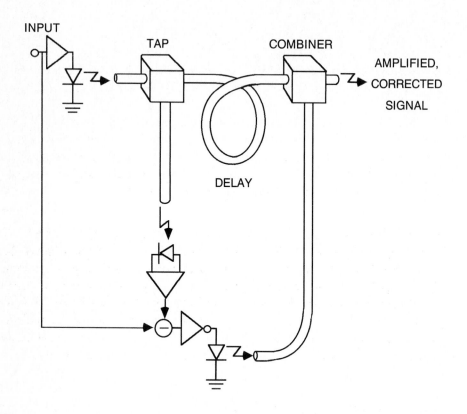

FIG. C-11. Light-domain feedforward system.

Real "design" entails choice of specific transistors, layout and interconnection. Since most such amplifiers will, in whole or in part, be integrated on a semiconductor chip, such choices are beyond the scope of this text. Suffice it, then, to simply choose resistor values on first order bases, and assume the transistors will have reasonable parameters.

The voltage gain of the first stage, given by $\dfrac{R_1}{r_e}$ at DC (where R_1 is the collector load resistor of the first transistor), determines the open-loop gain of the circuit, and should be large in order to assure approximately operational characteristics. R_1 should therefore be large, but not so large as to compromise

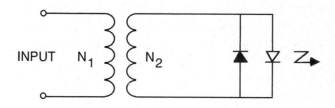

FIG. C-12. Alternative driver.

the ability to drive the second stage, whose impedance, fortunately, is readily made large, being given at DC by $\beta_0 R_2$ (where R_2 is the emitter load of the second transistor). R_2 should be chosen to be small enough to provide a low output impedance to the second stage, and high enough to contribute to a large input impedance. R_f (the feedback resistor) should be chosen to be as high as possible to contribute to a substantial transimpedance and to keep its thermal noise contribution within bounds but as low as possible for high-frequency operation. Let us choose $R_1 = 5,000$ ohms, $R_2 = 100$ ohms and $R_f = 10,000$ ohms.

Then the rms noise current contribution, given by

$$\sqrt{\frac{4kTB}{R_f}},$$

assuming a 100 MHz bandwidth and 20° C (293° K), becomes

$$\sqrt{4\times1.38\times10^{-23}\times293\times100\times10^6\times10^{-4}} = 0.0127 \ \mu A.$$

The nominal transimpedance of the amplifier is $R_f = 10,000$ ohms.

Note that, with an incident optical power of 1 μW, and a Responsivity of about 1, the current signal at the input of the amplifier is about 1 μA. The output voltage of the amplifier would then be about 10 mV.

4-8 Integrating Amplifier

An integrating amplifier is somewhat simpler in makeup than a transimpedance amplifier, comprising a shunt resistor at the input and a common emitter (assuming bipolar) transistor configuration. The shunt resistor should be as large as possible to reduce its noise contribution, but the amplifier's dynamic range is reduced as this resistor grows. (The frequency response effect can, of course, be compensated by equalization later in the amplifier chain.)

The value of the collector resistor is similarly constrained by the gain allowable in order to maintain adequate dynamic range.

Let us choose a value of 1,000 ohms for the collector resistor and 10,000 ohms for the input shunt resistor. Then the noise contribution of the shunt resistor is identical to that of the feedback resistor for the transimpedance amplifier discussed in the preceding exercise. The gain of the amplifier would, depending on the transistor and the frequency chosen, be in the neighborhood of 100 or more. To a first approximation, the 1 μw incident signal discussed above would result in a 0.01 volt signal at the input, and a signal of the order of 1.0 volt would result at the output.

4-9 Dynamic Range

The dynamic range of a preamplifier would have to be of the order of the loss budget for the fiber portion of the system if it were to operate as well when butted against the source as when a maximum length of fiber is inserted. This would translate to a range of some 30 to 70 dB, and would clearly be beyond the range of a high-impedance amplifier. A carefully designed transimpedance amplifier would be necessary.

4-10 Clock-Recovery Need

Clock recovery circuitry is unnecessary for any modulation scheme that is not digital, e.g., FM, PDM, PPM. It is also unnecessary for digital systems where a clock signal is supplied separately, e.g., intra-computer links, some encrypted data links, etc.

CHAPTER 5 EXERCISES

5-1 Square-Wave FM

The rise and fall times of lightwave signals often differ (the trailing edge typically being slower). Using both edges (as opposed to the leading edge only) in such cases tends to increase jitter which translates to noise in the recovered signal.

5-2 Best S/N Ratio

It will be recalled that the ratio $\dfrac{(S/N)_0}{(S/N)_i}$ (which will be called I below) is given by:

$$\text{PWM: } 1/8 \ (\frac{B}{f_m})^2$$

$$\text{FM: } 3 \ (\frac{\Delta f}{f_m})^3$$

$$\text{PCM: } 2^{2n}.$$

For FM, applying Carson's rule, $B \geqslant 2(f_m + \Delta f)$. Taking equality, $\Delta f = B/2 - f_m$. Then,

$$I = 3 \left[\frac{B/2 - f_m}{f_m} \right]^3 = 3 \left[\frac{B - 2f_m}{2f_m} \right]^3 .$$

Suppose $B \gg 2f_m$, then $I \approx 3/8 \ (\frac{B}{f_m})^3$.

More typically, $B \approx 4 f_m$, and $I \approx 3$. This can be compared to PWM, which, under the same conditions, produces an $I \approx 2$.

For PCM, sampling must take place at a rate $\geqslant 2f_m$, and uses n bits per sample, for a rate $\geqslant 2nf_m$ bps. Choose equality for economy.

Note also that with NRZ encoding, B = one-half the bit rate or nf_m, and $n = \dfrac{B}{f_m}$.

I then is given by $2^{\frac{2B}{f_m}}$.

For $B = f_m$, the ratio is 4, but this would correspond to only 1 bit per sample.

More typically, 8 bits are used per sample, which corresponds to $B = 8f_m$ for which $I \approx 64,000$.

For the same value of B, for PWM, $I = 8$, and for FM, $I = 81$.

5-3 Modulators and Demodulators

All of the modulation schemes need a low-pass filter as part of the demodulation process; this element will therefore be assumed and not shown.

a. A very simple pulse-width modulator can be built as indicated in Figure C-13, which makes use of the stored charge in the transistor when it is saturated and then cut-off so rapidly that a transient condition of emitter-base diode cut-off with the collector-base diode still forward biased is induced. The modulating signal can be coupled into the collector via a transformer, and an emitter-follower can be used as an isolation stage to prevent loading of the relatively high-impedance collector.

The bottom of the figure depicts the voltages at three points as might be viewed on an oscilloscope. The sampling pulse stream would be portrayed as indicated, the voltage at the collector of the first transistor would display the peculiar behavior of such a device used as indicated, and the slicer would produce the pulse-width modulated output.

A low-pass filter is sufficient for demodulation.

b. A classical FM modulator employs a variable reactance element to alter the operating frequency of an oscillator in response to the amplitude of the modulating signal. Figure C-14 illustrates such a circuit, where the voltage-variable capacitance of a reverse-biased diode is employed. (It is important that the

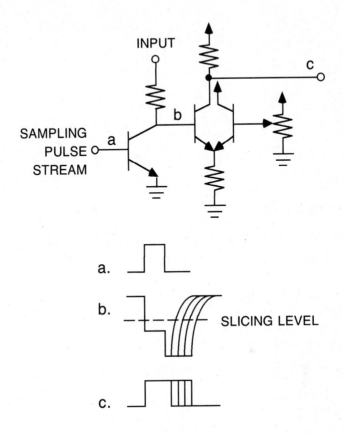

FIG. C-13. Pulse-width modulating circuit.

drive circuit to provide a bias to keep the diode reversed in this implementation).

Demodulation can take on a number of forms: simple slope detection, ratio detection, etc. Always, of course, followed by a low-pass filter. The implementation shown is of a balanced discriminator preceded by a limiter.

c. Pulse-Code Modulation comprises a sample and hold circuit followed by a quantizer and an encoder (see Figure C-15). (The details of an implementation are sufficiently messy to justify settling for block diagrams.)

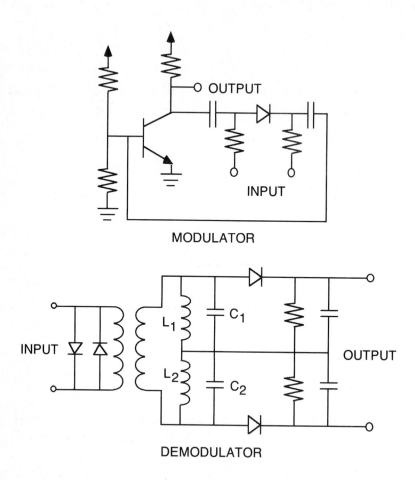

FIG. C-14. FM modulator and demodulator.

The demodulator requires means for clock recovery, a digital to analog converter, and a low-pass filter.

The complexity increases from PWM to FM to PCM. The relative costs of the elements are more difficult to determine, however, because these are tied to the level of integration (digital components can be extremely highly integrated), the volume of production (which can change with time), and other considerations.

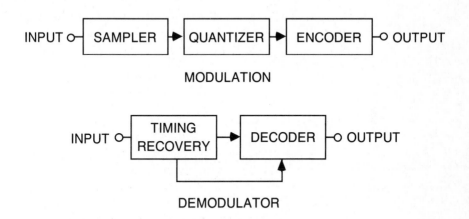

FIG. C-15. Pulse-code modulator.

5-4 Dual Modulation

The separation under ideal conditions is not difficult. The incoming signal can be split down two circuit paths: one to detect the amplitude modulation, the other to detect the width modulation.

The width modulation detector need only apply limiting to eliminate amplitude variation, then a low-pass filter will recover the width-modulated signal.

The amplitude modulation detector must first strobe the incoming pulses to eliminate the width variation. A strobe signal beginning at the rising edge of the incoming pulse with a duration calculated to be shorter than the pulse produced by the most extreme depth of width variation can perform this function. The resulting strobed signal can then be submitted to a low-pass filter for recovery of the amplitude-modulated signal.

5-5 PCM S/N

The gap between the improvement ratios for FM corresponds to the onset of thresholding, above which the performance is excellent, and below which it is abysmal. The threshold area corresponds to performance marked by significant degradation.

5-6 Differential PCM

The synchronization pulses are the most difficult signals to accurately convey with a delta modulation system. They represent the sharpest rate of rise of any signal that can be encountered, while their temporal position is extremely important for proper rendering of the signal at the receiver.

One potential means for handling this problem when the nature of the signal is known at both ends of the transmission link would be to run a clock-recovery circuit at the receiving end which, in effect, would anticipate the sync signals and square them up upon their arrival independent of how poorly simple delta modulation might otherwise portray them.

5-7 PPM from PWM

Figure C-16 depicts a circuit that could perform the required functions. The PWM signal is first differentiated, then rectified, then used to trigger a one-shot, producing a set of pulses whose position is proportional to the amplitude of the modulating signal at the time of sampling.

5-8 Carson's Rule

It will be recalled that Carson's Rule dictates that the necessary bandwidth to adequately convey an FM signal is given by

$$B \geqslant 2(f_m + \Delta_f) \ .$$

(Note that the carrier frequency is of no consequence.)

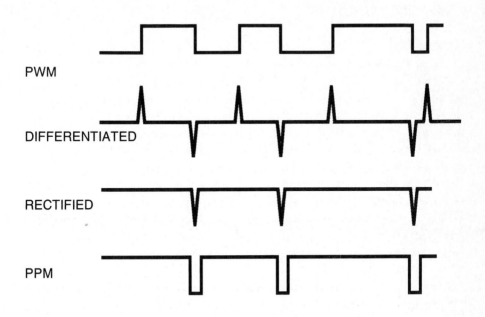

FIG. C-16 PWM to PPM conversion.

Clearly, then, the minimum B required in this case is

$$B = 2(4.5 + 10) = 29 \ MHz \ .$$

A "typical" fiber might have a bandwidth of 500 MHz-km and be one km in length. Then, the number of channels, assuming 1 MHz of guard bandwidth, would be $500/30 = 16$ channels.

In the real world, unless linearizing or spacing techniques are used, or some ruggedizing modulation approach (e.g., FM) is employed, it may be unwise to attempt to convey more than one TV channel over a fiber facility.

CHAPTER 6 EXERCISES

6-1 Loss Budget: Optoisolator

An optoisolator is a specialized piece of electronic equipment intended to provide electrical isolation between two elements of equipment to allow separate power and grounding for signal handling or to prevent the possibility of excessive voltages, or both. The distance between the transmitter and receiver is typically microscopic, so that it becomes a degenerate transmission system.

In real optoisolators, the detector is often a phototransistor, which provides good gain, but is generally much slower than photodiodes. For the purposes of this exercise, however, it will be assumed that the isolator is fashioned of the same componentry as the systems studied in the text.

Several of the elements of the loss budgets that were discussed in the text become null: the length (which is negligibly small), the attenuation of the transmission medium, the number of splices and connectors (both of which are 0), and the corresponding losses. All that remains is the emitted power, the coupling loss, the aging margin, the detector quantum efficiency, and the receiver sensitivity.

Assume that:

$$P_E = -10 \text{ dBm}$$
$$L_C = 3 \text{ dB}$$
$$A = 10 \text{ dB}$$
$$\eta = 0.5$$

Then the receiver sensitivity need be only -26 dBm.

6-2 60 MHz System

This is a situation very similar to one examined earlier in the chapter. 300 feet is about 100 meters, so that, in view of the frequency requirement (note, in Hz, not bps), the application can be accommodated with large-core fiber (e.g., 200

μm core) for easy coupling from an inexpensive surface-emitting LED. (60 MHz bandwidth is adequate for conveying domestic US television Channel 2 signals.) It should be noted that the transmitter and receiver would necessarily have to be able to accommodate analog signals.

$$P_E = -10 \text{ dBm}$$
$$L_C = 6 \text{ dB}$$
$$l = 0.1 \text{ km}$$
$$\alpha_f = 10 \text{ dB/km}$$
$$n = 0 \text{ Splices}$$
$$L_S = \text{(Not Applicable)}$$
$$m = 2 \text{ Connectors}$$
$$L_m = 1 \text{ dB/Connector}$$
$$A = 3 \text{ dB}$$

The receiver sensitivity would have to be -22 dBm. This value will provide an adequate S/N ratio without significant strain in receiver design.

6-3 200 Mbps System

The bit rate allows an LED and graded-index fiber (the bandwidth requirement would be 100 MHz×4 km). It will also be assumed that operation will be at short wavelength.

$$P_E = -10 \text{ dBm}$$
$$L_C = 6 \text{ dB}$$
$$l = 4 \text{ km}$$
$$\alpha_f = 3 \text{ dB/km}$$
$$n = 5 \text{ Splices}$$
$$L_S = 0.2 \text{ dB}$$
$$m = 4 \text{ Connectors}$$
$$L_m = 0.3 \text{ dB/Connector}$$
$$A = 5 \text{ dB.}$$

The receiver would require a sensitivity of -35.2 dBm.

6-4 1 Gbps System

This application would require an ILD because of the high-frequency operation, and single-longitudinal mode capability because of the length of the span.
Single-mode fiber operated at minimum attenuation and at or near a dispersion
zero would be required. The bandwidth-distance product needed would be 50
GHz-km.

$$P_E = 0 \text{ dBm}$$
$$L_C = 3 \text{ dB}$$
$$l = 100 \text{ km}$$
$$\alpha_f = 0.15 \text{ dB/km}$$
$$n = 20 \text{ Splices}$$
$$L_S = 0.1 \text{ dB}$$
$$m = 4 \text{ Connectors}$$
$$L_m = 0.2 \text{ dB/Connector}$$
$$A = 5 \text{ dB.}$$

The receiver sensitivity required would be -25.8 dBm, which is about right for a
BER of 10^{-9} at 1 Gbps. (Note that a 1.7 Gbps long-haul system has been
announced.)

6-5 12,000 Mile System

Clearly the indicated system could not be accommodated by a single span, so
some assumptions will have to be made. (It will be noted that 12,000 miles is
roughly half-way round the world.) If 100 GHz-km fiber were available
(assuming a single-mode ILD), then a 400 km span would be possible if
bandwidth were the only limitation. Subject to assumptions below, it will be
seen that attenuation is the stronger limit and would pull the span down to 150
km.

$$P_E = +10 \text{ dBm}$$
$$L_C = 3 \text{ dB}$$
$$l = 150 \text{ km}$$
$$\alpha_f = 0.2 \text{ dB/km}$$
$$n = 30 \text{ Splices}$$

$$L_S = 0.1 \text{ dB}$$
$$m = 0 \text{ Connectors}$$
$$L_m = \text{(Not Applicable)}$$
$$A = 10 \text{ dB.}$$

The receiver sensitivity required would then be -36 dBm. 128 repeater spans would then be required.

Of course, if ultra-low-loss fiber becomes available and/or coherent reception techniques could be utilized, the number of spans could be significantly reduced.

6-6 Highest Channel Capacity System

This exercise requires speculation. Suppose that it were possible to approach the quantum limit via coherent detection, use an arbitrarily spectrally pure ILD output, operate at a wavelength of minimum attenuation and zero material dispersion, and have a fiber with the attenuation promised of new materials. Further, the attractive assumption could be made that the loss associated with splices is comparable to that of the fiber itself between splices.

First, it will be assumed that coherent detection techniques enable us to approach arbitrarily close to the quantum limit, so that we may use the expression:

$$BER = e^{-N},$$

where, it will be recalled, N is the number of photons necessary to represent a "one" in order to obtain the indicated bit error rate.

We can also write an expression for the average received power:

$$P_R = N \times h\nu \times (B/2)$$

Substituting from the previous expression to eliminate N,

$$P_R = -ln(BER) \times h\nu \times (B/2)$$

(It will be noted that the minus sign will be negated because the BER will have a negative exponent.)

Solving for B,

$$B = \frac{2\lambda \times P_R}{hc \times \ln(\frac{1}{BER})}$$

Now, assuming that the BER is 10^{-9}, the wavelength is 1 μm, and the received power is -50 dBm, we find that the B is

$$B \approx 5 \times 10^9.$$

The span length, assuming a launched power of 100 mW and an attenuation of 0.001 dB/km, and taking into account the assumed splice loss, would be 35,000 km (21,875 miles), assuming no margins.

6-7 ILD Sparing

"Invention" implies that any number of schemes may surface. One that is both practical and practicable will be described.

The TAT-8 submarine cable system is presently planning to employ a sparing technique for its lasers that will mechanically switch fibers using a linear stepped movement to align an outgoing fiber with a succession of alternative ILD-driven fibers as successive failures occur.

6-8 Field Trial Critique

A critique must necessarily have an ideal with which to compare, and the identification of an ideal is somewhat subjective. Rather than proposing an ideal from a technical viewpoint, it might be more reasonable to base the critique on the state of the art at the time of the trial's deployment and perhaps indicate how some things would be done differently today.

The Japanese HI-OVIS system need apologize to no one because it was the first significant effort to apply fiber to the residence. The use of two fibers was doubtless made because there was not an economically justifiable means for two-way transmission on a single fiber at the time. The maximum length was short because the system operated at short wavelengths and fiber attenuation was still on the high side, and/or the system design could be relaxed under the nondemanding loss budget constraints. The use of plastic cladding over an ungraded glass core was a reasonable economic choice because the cladding can double as the jacket, and the core can be economically produced of uniform material. Generally, the loss of such fiber is larger than for an all-silica fiber, but the application is undemanding.

The Yokosuka field trial experimented with a far broader scope of optical technology than HI-OVIS, including wavelength division multiplexing and duplex use of single fibers. Further, the range of services was much broader, encompassing the version of ISDN services codified at that time.

The Biarritz trial is interesting because of its choice of frequency division multiplexing (a virtually unique choice) and both its original and planned scope. The number of customers intended to be served ranks with BIGFON as far larger than for any other presently proposed project.

The BIGFON trial is remarkable because it entails the efforts of a multiplicity of domestic (German) manufacturers with designs that are independent yet must function cooperatively. The philosophy of using a 140 Mbps rate for transmission and switching of video is also quite different from domestic US thinking, which tends toward conserving bandwidth.

The Milton-Keynes FIBREVISION project chose pulse frequency modulation for the sake of economy (the modulators and demodulators are much less expensive than are the digital equivalents). The use of copper-clad strength members for electrical domain up-stream signaling is also worthy of note.

The Yorkville experiment's choice of analog transmission was appropriate to the era, but limited the extent of useful employment of fiber bandwidth.

The proposals for the Belgian system are particularly interesting because of the very large bandwidth to the customer (1.1 Gbps) downstream. Such bandwidth requires single-mode fiber, and the influence this data rate may have on international thinking could be considerable.

6-9 ISDN System

Recognizing that the augmented (Broadband) ISDN services include voice, data and video to and from the subscriber, a number of implementation choices are possible. Economics appears to be the overriding consideration, however.

Philosophically, the most elegant technique would use only a single fiber for transmission in both directions of all signals. Economics might dictate that an intermediate node be utilized that would have a single feeder route from the central office, and individual loop fibers to residential and business customers.

By the time such a system would be broadly available, digital television sets may have become common, and an arrangement whereby the digitally transmitted video could be directly processed within the television set without the expense and harm to the signal of converting back to base-band video, modulating a VHF or UHF carrier, then tuning and downconverting within the TV set may be reasonable.

Single-mode fiber to the customer could provide the early bandwidth needs and the capacity for future services.

CHAPTER 7 EXERCISES

7-1 Wavelength-Division Multiplexing

If each such combination and splitting actually exacted a 3 dB penalty, then each additional wavelength above the first would account for 6 dB of the budget. If 30 dB were available because of negligible fiber loss, then 6 wavelengths could be accommodated. Fortunately, it is possible to couple and decouple with considerably less than a 3 dB penalty.

7-2 Maximum Fiber-Optic Loop Lengths

Such a system would be analogous to a pair of telephone stations served by the same central office and at the limit of the allowed loop length (determined by the limits of the capability for adequate transmission, alerting, and detection of switch-hook status). For a fiber optics system, the overriding requirement would be attenuation limitations, including losses in the switch, which might be significantly greater than for a metallic equivalent.

Consider such a system comprised of two fiber spans emanating from a centralized light-domain switch. Setting up a loss budget, but assuming a receiver sensitivity and introducing a switch-loss component, it is possible to solve for l:

$$P_E = 7 \text{ dBm}$$
$$L_C = 6 \text{ dB}$$
$$l = ?$$
$$\alpha_f = 3 \text{ dB/km}$$
$$n = 0$$
$$L_S = (\text{Not Applicable})$$
$$m = 4$$
$$L_m = 1 \text{ dB}$$
$$\eta \approx 1$$
$$A = 5 \text{ dB}$$
$$L_{SWITCH} = 3 \text{ dB/stage}$$
$$P_R = -30 \text{ dBm}$$

Then, assuming 4 stages of switching, 10 dB remains to be allocated to the spans, and at the assumed 3 dB attenuation per km, the loop length (recognizing that there are two involved) would be 1.65 km, or approximately 1 mile. If lower-loss fiber were used, the length would be correspondingly greater.

7-3 Optical Switching Computer

The notion of an optical computer has been a dream for many years. Two possible implementation types are possible: electrooptical and strictly optical. Electrooptical devices, as presently envisioned, would employ devices which would require dimensions spanning several wavelengths, inevitably rendering them physically larger than purely electronic logic devices, which already have feature sizes that are a fraction of a wavelength of typically employed light. Relatively large voltages are also employed, applied to relatively large electrodes, making a combination of large C and large V swing, necessitating high currents for high speed. Pure photonic devices hold more promise for speed, but it should be recognized that light signal propagation rates are not very much faster than that of electrical signals. Further, dissipation may present a larger problem than for electrical circuits. Pure photonic circuits may lend themselves best to parallel computation.

7-4 Optical Logic Circuits

In an electrooptical system, the optical output is a function of electrical inputs. Viewing the electrical inputs as binary variables, two couplers in series would function as an "AND" gate, two in parallel as an "OR" gate, and complementation could be performed by taking the alternative coupler output.

7-5 Fiber-Sensing Parameters

Any phenomenon that could cause a detectable change in the output of a fiber system is a candidate for fiber sensing: temperature, pressure, acceleration, magnetic field intensity (with the fiber coated with or bonded to a magnetostrictive material), fluid level, acoustic power, displacement, torque, strain, radiation, flow rate, pH, blood oximetry, etc.

7-6 Integrated Optics Light Source

For a fully integrated optical system, a distributed feedback ILD internal to the substrate would approach the ideal in that it could be fabricated at any point in the substrate (as opposed to necessarily being at a facet).

7-7 Fiber-Guided Torpedo

It would appear to be important to utilize a very small cabled fiber for this application to keep its bulk minimal. The payout mechanism might best be similar to that of a spin-casting reel as used for fishing, minimizing the stress upon the fiber. Both a one-way and a two-way communication mechanism could be visualized. The one-way approach would have to incorporate means for the controller on the submarine to determine the position of the torpedo at any time so that intelligent course-change messages could be dispatched to it. The two-way mechanism would allow the torpedo to send information concerning its actual (as opposed to assumed) response to control signals, and would yield better performance.

The one-way implementation is reasonably self-evident. The two-way case could probably be most easily and economically handled using two wavelengths for easy separation.

APPENDIX A EXERCISES

A-1 Laboratory Plan

The answer to this is "free-style" in the sense that there are as many different answers as there are people willing to do the planning.

A-2 Time-Domain Reflectometer Response

See Figure C-17. The positive spikes are all due to reflections at the interfaces. The attenuation of the single-mode fiber is likely less than that of the graded-index fiber because the dopant level is lower (unless there is a mismatch in spectral response). The profound loss at the point of interface between the two fiber types is due to the very poor coupling back into the single-mode fiber from the graded-index fiber. (Note that the trace therefore does not accurately portray the power level in the graded-index fiber because the coupling from the single-mode to the graded-index fiber should be quite good.)

A-3 Technician Background

Although an Electronics background would find some utility, a strong Physics background might prove more valuable. (A blend of the two could be ideal.)

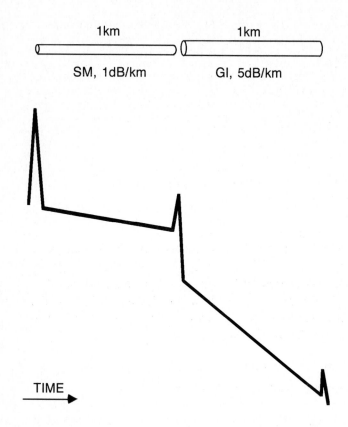

FIG. C-17. Exercise A-2: Time-domain reflectometer response.

GLOSSARY

a: Fiber core radius.

Acceptance angle: The angle of incidence relative to a normal to the entrance surface below which light will be guided within an optical fiber.

ADM: Adaptive Delta Modulation. A technique intended to overcome the slope-overload problem of simple Delta Modulation by changing the increment size of the demodulated signal when a succession of like changes is encountered.

Alpha-decay: A nuclear decay process which emits the nucleus of a helium atom or a doubly ionized helium atom.

Angstrom: Unit of measure equal to 10^{-10} meters.

Antineutrino: An electrically neutral particle of very small rest mass, spin quantum number of ½, and spin oriented parallel to the particle's linear momentum.

Antireflection Coating: A coating of one or more layers that reduces the reflectivity of a surface.

Attenuation: The reduction in amplitude of a signal. Usually expressed as a power ratio in dB or Nepers.

Avalanche Photodiode (APD): A photodiode which in use is biased close to the avalanche breakdown potential. Captured incident photons result in the production via an avalanche process of more than a single hole-electron pair.

B: Magnetic flux density. Units of webers per square meter. Also Bandwidth in units of Hz.

Bait-Rod: The rod used in the OVD or VAD process to begin deposition upon.

Bandgap: The least energy difference between the conduction and valence bands; symbol E_g. Usually expressed in electron-volts.

Bandwidth: The frequency band between the half power points; symbol B; units Hz. Electrical value $= 1/\sqrt{2}$ of optical value.

Baseband: The frequency band occupied by an unmodified signal, before it is used to modulate some carrier. Examples: NTSC television occupies about 4.5 MHz baseband, while a telephone voice signal occupies about 4 kHz baseband.

BER: Bit Error Rate. A measure of the reliability of digital transmission. A typical goal is 10^{-9}.

Bessel Function: A set of periodic functions which are solutions to a class of second order ordinary differential equations.

Beta-decay: A nuclear decay process which emits an electron and an antineutrino.

Binomial Distribution: A distribution assuming discrete arrival statistics. In the limit as discrete time intervals tend to zero, it approaches a Poisson distribution.

BISDN: Broadband Integrated Services Digital Network (Sometimes written ISDNB) - A set of international standards for telecommunication that includes voice, data, and broadband video.

Burrus Diode: A type of surface-emitting LED.

Butt Coupling: Coupling of two fibers via butting together prepared ends.

Carson's Rule: Rule giving a lower bound on the bandwidth necessary to adequately convey an FM signal: the bandwidth should be greater than or equal to twice the sum of the modulating frequency and the maximum deviation of the carrier.

Cascode Amplifier: An amplifier dating back to vacuum-tube days. The Miller effect is avoided by coupling the collector (or drain or plate) of a first transistor to the emitter (or source or cathode) of another, so that the voltage swing of that point is virtually nil.

CATV: Community Antenna Television. Essentially cable distributed television whether or not there is a shared antenna. The name arose from early applications where small valley towns were unable to receive television until a shared antenna was erected on a nearby hill and the signal was distributed via cable. The appellation persists for any guided transmission of commercial TV programming.

Caustic Surface: The surface bounding the innermost excursion of a guided skew ray in a step-index fiber, and the inner and outer excursions of bound mode rays for graded-index fiber.

CCITT: An international standards-setting body.

C^3 *Laser:* Cleaved-Coupled-Cavity Laser. A laser comprising two distinct cavities coupled via a short air space. Typically only one of the structures is modulated; the other may, however, be powered on a DC basis.

Central Dip: The pronounced dip in index of refraction often seen in fiber prepared via the MCVD or OVD process as an artifact of the originally hollow preform.

Cerenkov (sometimes spelled Cherenkov) radiation: Radiation emitted when charged particles enter a medium where their velocity exceeds the phase velocity of light in that medium.

Cladding: The material surrounding the core of an optical fiber. Made to have an index of refraction lower than the core (step-index) or matching the peripheral limit of the core (graded-index). Serves in single-mode fiber to shield guided power from the lossy jacket.

Clamping: Also called limiting. The limiting of a voltage signal via a diode to a fixed reference voltage (+ the diode drop).

CML: Current-Mode logic. See ECL.

Coherence: The property of field components at different points in space being phase related.. Alternatively stated, displaying spatial or temporal correlation.

Collector Compensation: A technique for extending the bandwidth of a bipolar transistor amplifier by placing reactive elements in the collector whose greater impedance at elevated frequencies will support the gain of the amplifier at those frequencies. Analogous to plate compensation for vacuum tube circuits and drain compensation for FET transistor circuits.

Companding: A technique for improving the performance of a modulation technique by predistorting a signal before modulation and postdistorting in a complementary manner after demodulation. The signal is typically compressed in range before modulation and expanded after demodulation.

Conduction Band: The band of energy levels in a semiconductor above the forbidden band, where excited electrons exist until some mechanism brings about recombination and reversion to the valence band or a trapping state.

Core: The usually cylindrical innermost constituent of an optical fiber. Made to have a higher index of refraction than its cladding in order to promote guidance.

Critical Angle: The angle (to the normal) of incidence of light upon a boundary between media of differing refractive indices above which total reflection occurs.

Cutoff Wavelength: That wavelength at which $W = 0$, which corresponds to $\beta = n_2 k$. Above this wavelength, a mode becomes lossy.

D: Electric flux density. Units of coulombs per square meter.

Dark Current: The leakage current in a detecting device when no signal is incident (i.e., the source is dark). A noise contributor. Higher, for example, in germanium than in silicon.

dBm: A power level expressed in dB referenced to one milliwatt.

DC: Double Crucible technique: A technique for manufacturing fiber which employs two, concentric crucibles containing molten glass, with apertures at their bottoms through which fiber may be drawn with a core comprised of material from the inner crucible, and cladding comprised of material from the outer crucible.

Decibel (dB): A logarithmic ratio of power levels: the value in dB is equal to 10 times the base-ten logarithm of the ratio of the two powers of interest.

Del (∇): A differential vector operator defined for Cartesian coordinates as $\frac{\partial}{\partial x}\mathbf{i} + \frac{\partial}{\partial y}\mathbf{j} + \frac{\partial}{\partial z}\mathbf{k}$.

Del Squared (∇^2): The Laplacian operator; the divergence of the gradient.

Delta Modulation: A technique which transmits only the change in value from the last sample as opposed to the absolute value. In practice, the direction of change is indicated by a single bit.

Delta sub f (Δ_f): Maximum frequency deviation of the carrier of an FM signal.

Deviation: The maximum amount (in Hz) by which the carrier frequency in an FM system is deviated from its nominal value.

DFB Laser: Distributed Feedback Laser. An ILD that utilizes periodic structures instead of facet mirrors to contain light and promote single longitudinal mode operation within the active cavity.

Dichroic Mirror: A surface that is transmissive at one wavelength and reflective at another.

Differential Gain: The difference between unity and the ratio of the output amplitudes of a small high-frequency sinewave signal at two stated levels of a low frequency signal on which it is superimposed. A figure of merit for transmission of video signals.

Differential Phase: The difference in output phase of a small high-frequency sinewave signal at two stated levels of a low-frequency signal on which it is superimposed. A figure of merit for transmission of video signals.

Direct Bandgap: Property of a material such as GaAs which, for an electron transition from the conduction to the valence band, conserves both energy and momentum.

Discriminator Circuit: A decision circuit used to determine whether a signal at a given time represents a digital 1 or a 0.

Dispersion: The temporal spread of a signal due to modal, chromatic or other effects. Expressed in ns per nm of wavelength per km.

> *Modal:* Dispersion due to the differing flight times of photons following an axial path and those following the most extreme path, corresponding to entry at the acceptance angle.

> *Material (or Chromatic):* Dispersion due to the differences in refractive index as a function of wavelength.

> *Waveguide:* The delay distortion due to frequency dependency of the mode group velocity.

Distribution Facility: The transmission link between the extremity of a feeder facility and the drop.

Drop: The transmission link from the extremity of a distribution facility (e.g., a pole or distribution box) to the customer. In a telephone installation, it would be the wire pair between a residence and the nearest serving area interface.

Dynamic Range: The range of input level presented to an amplifier over which undistorted amplification will take place.

Dynode: An electrode within a photomultiplier tube, usually at an intermediate potential relative to the cathode and plate, which is coated with a material which enhances the coefficient of secondary emission. Typically, multiplication takes place at each of a succession of dynodes.

E: Electric field intensity. Units of volts per meter.

ECL: Emitter-Coupled Logic. A form of current-mode logic utilizing two transistors whose emitters share a common (for npn) current sink. A high-speed, high-dissipation circuit type.

Eigenvalue: One of a set of discrete solutions to certain classes of equations.

Electron-volt: The energy acquired by an electron in falling through a potential of one volt. One electron-volt is equal to 1.6020×10^{-19} joule.

Electro-Optic Effect: An effect evidenced by materials such as Lithium Niobate ($LiNbO_3$) which causes the index of refraction to change in the presence of an electric field.

Emitter Compensation: A method used in association with bipolar transistor amplifiers which places a reactive element in the emitter whose lowered impedance at higher frequencies supports the gain of the amplifier at those frequencies. Analogous to cathode compensation in vacuum tube circuits and source compensation in FET transistor circuits.

Epitaxy: The process of growing new material onto a host material while preserving the crystallographic structure.

Epsilon (ϵ): Permittivity. Units of farads per meter.

Equalization: The process of compensating for the bandwidth limitations of an amplifier after the fact using a filter with compensating characteristics. Virtually required following an integrating amplifier.

Eta: Quantum efficiency for a detector (η): the ratio of usefully captured to incident photons.

Evanescent Field: The nonpropagating field that exists in the cladding of a fiber conveying power in its core.

Exclusive-Or: Logic function of two variables given by $A\bar{B} + \bar{A}B$.

Extinction Coefficient: The imaginary component of a complex index of refraction.

Extinction Ratio: The ratio of the power level representing a mark to that representing a space. In general, the larger, the better.

f_m: The maximum frequency of a modulating signal.

Fabry-Perot Resonator: An electromagnetically resonant cavity comprised of an active region bounded by two mirrors.

Feeder: The transmission link between a central office (or head end) and a remote node or branch point.

FET: Field Effect Transistor; a transistor utilizing only one species of carrier (holes or electrons).

Flash Encoder: An encoder that strives for rapid operation by utilizing a multiplicity of encoding elements in parallel.

Flicker Noise: A 1/f type noise which is important only at relatively low frequencies.

Fluence: The time integral of radiation flux expressed in particles per square cm or energy per square cm.

FM: Frequency Modulation. A technique for varying the instantaneous frequency of a carrier signal about a nominal value in proportion to the amplitude of a modulating signal.

Forbidden Band: The energy band between the valence band and the conduction band.

Fresnel Reflection: Reflection phenomenon at an interface between materials of differing indices of refraction; constituents calculable from the Fresnel reflection equations.

Fusion Splicing: A splicing technique which melts (fuses) the prepared fiber ends together. Electric arc typically used.

Gamma ray: Energetic photonic radiation occurring when a nucleus decays from an excited state to its ground state. Energies range from 0.1 to 10 MeV.

GHz: Giga-Hertz; 10^9 Hertz.

Graded-Index Fiber: Fiber whose core has a refractive index that varies with the radius.

Grating (Diffraction): A surface with closely-spaced ruling or grooves. Historically produced using high-precision ruling lathes. Gratings are often replicated from mechanically ruled versions (e.g., gold replication).

GRIN-Rod: A rod of graded-index glass several times the diameter of typical fiber. Used, e.g., for combining and splitting signals. Typically manufactured utilizing an ion-exchange process.

H: Magnetic field intensity or magnetomotive force. Units of amperes per meter.

h: Planck's constant: 6.626×10^{-34} J/Hz.

Hankel Functions: Linear Combinations of Bessel Functions t⊦ are also solutions to Bessel's Equation.

Hard Limit: Clamp.

Head-end: The point of origin of transmⁱⁱeld of telephony. of programming for a CATV system. Analogous to the central office ⸴

Helical Ray: A light ray whicᵗ ɹes a helical path at constant radius in a fiber. A special case of a sᵗ

Heterodyne: A demodulation technique employing one or more frequency conversions. The local oscillator differs in frequency from the carrier.

Heterojunction: A pn junction utilizing differing materials on either side.

High-Impedance Amplifier: See Integrating Amplifier.

Homodyne: A single-conversion demodulation technique (no intermediate conversions). A local oscillator running at the frequency of the carrier is beat against the incoming signal to recover the modulating information.

Homojunction: A pn junction utilizing the same material on both sides.

Hybrid Mode: A mode with axial components of both E and H fields.

Hydroxyl Ion: The OH^- ion, often referred to simply as water. A potential major cause of attenuation in high-silica fiber.

ILD: Injection Laser Diode. A semiconductor device which injects carriers across a pn junction into a cavity capable of supporting stimulated emission.

Index of Refraction: The ratio of the speed of light in a vacuum to the speed of light in the medium of interest. Always greater than or equal to 1. May be complex.

Indirect Bandgap: Property of materials such as silicon and germanium for which the transition from the conduction band to the valence band of an electron is not both energy and momentum conserving, requiring the generation of a phonon as well.

Insertion Loss: The loss resulting from the insertion of an additional element into a transmission path.

Integrating Amplifier: An amplifier without a feedforward or feedback equalizer. It has a poor dynamic mechanism which typically narrow bandwidth requiring a following times called a *High-Impedance* amp but good noise characteristics. Some-

Intermodulation Distortion Products: The distorting spectral components produced by nonlinear mixing of signals.

ISDN: Integrated Services Digital Network - A set of international standards for digitally providing voice, data and video services.

J: Current density. Units of amperes per square meter.

Jacket: Protective covering placed immediately over the cladding of an optical fiber for protection against abrasion and moisture intrusion.

Lambda (λ): Wavelength; usually specified as free-space value.

Lambertian Radiation Pattern: A radiation pattern characterized by a power level at any external point proportional to the cosine of the angle relative to a normal to the radiating surface.

Laser: Acronym for Light Amplification through Stimulated Emission of Radiation. Semiconductor types are *Injection* Lasers.

Lasing: The process of producing coherent light via stimulated emission.

LED: Light Emitting Diode.

Liquid Phase Epitaxy: The process of epitaxial growth from material in the liquid phase (LPE). Usually performed with a movable boat containing the molten material.

Manchester Coding: One of a class of encoding techniques designed to guarantee a minimum pulse rate. This assures that phase-locked mechanisms see pulses often enough to retain synchronism and that AC-coupled amplifiers do not have recovery problems.

Mark: The presence of a signal intended to represent a logical 1.

Maser: Acronym for Microwave Amplification through Stimulated Emission of Radiation.

Meridional Ray: A light ray whose path is in a plane containing the axis of the fiber core.

Metal-Organic Epitaxy (MOE): A technique of epitaxial growth where the constituents to be deposited are in the form of an organic compound, the unwanted portions of which dissipate in the deposition process.

MCVD: Modified Chemical Vapor Deposition. A technique of fiber manufacture which deposits, via a vapor oxidation process, doped silica upon the inner surface of a rotating cylinder of high-quality glass.

MHz: MegaHertz; 10^6 Hertz.

Microbend; A microscopic bend or kink that occurs in fiber during the cabling and handling processes.

Microphonic: Physical vibrations influencing a signal being transmitted or amplified.

Micron: One millionth of a meter (micrometer or μm).

Miller Effect: The effect which multiplies the collector-base capacitance of a transistor by the gain of that stage; the effective capacitance is known as the Miller Capacitance.

Mixing: The process of combining two or more signals impressed upon a non-linear element to produce a distorted resultant resolvable (in general) into spectral components including the original signals, harmonics, and sum and difference signals.

Modal Noise: Noise introduced into a light signal due to changes in the speckle pattern at one or more points in the transmission path. The speckle pattern is the result of the use of a coherent light source. Since a significant fraction of the signal power is assignable to each speckle, any mechanical movement that would bring about the loss of a speck would cause a corresponding reduction in the delivered power.

Mode: Principally the propagating electromagnetic modes which exhibit distinct physical configurations, and represent the discrete solutions to the field equations subject to physical boundary conditions.

Mode Field Radius: The distance from the axis of a single-mode fiber at which the (assumed) Gaussian Field is $1/e^2$ of its axial intensity. For a non-Gaussian field, a far-field second moment definition is used.

Modified Bessel Functions: Solutions to the Modified Bessel Equation which differs from the Bessel Equation in that the argument is imaginary. These functions are nonoscillatory and decaying and are suitable for solutions to the wave equations in the cladding region.

Modulation: The process of impressing the information from a modulating signal onto a carrier or into a pulse regime.

Molecular Beam Epitaxy: (MBE) The process of epitaxial growth from a beam of molecules directed at a target in a vacuum chamber. Relatively low rate of growth, but excellent material purity.

Mu (μ): Permeability. Units of henrys per meter.

Multimode Fiber: Fiber capable of supporting a multiplicity of bound modes at the wavelength of transmitted light. May be either graded index or step index.

Neper: A logarithmic ratio: the value in nepers is the natural logarithm of the ratio between the two levels of interest.

Neutrino: An electrically neutral particle of very small rest mass and of spin quantum number ½. The spin is antiparallel to the linear momentum of the particle.

NRZ: Non-Return to Zero. An encoding technique which reduces by half the bandwidth required of a transmission channel by not requiring a signal to return to a zero level between successive ones.

NTSC: National Television Standards Committee - a standards committee that specified the format for domestic US television. (This standard has been adopted in many other parts of the world as well.) This standard utilizes 525 horizontal scan lines. Other standards include PAL and SECAM.

Nu (ν): Frequency in Hz or mode number (dimensionless).

Numerical Aperture: The sine of the acceptance angle. A figure of merit relative to the ability to accept offered light for guidance.

Nyquist Criterion: A criterion that specifies the sampling rate that must be used in order to assure alias-free signal recovery - the sampling rate must equal or exceed twice the highest frequency component of the band-limited signal of interest.

OVD: Outside Vapor Deposition. A technique for fiber manufacture that deposits via a flame oxidation process, doped silica upon the outside of a rotating cylinder.

PAL: Phase-Alternating Lines - A television standard that has been widely adopted in Europe which alternates the sense of the chrominance phase reference between successive lines. 625 horizontal scan lines are used in this system.

PAM: Pulse Amplitude Modulation - A modulation technique which represents the amplitude of a signal sample with a pulse whose amplitude is proportional to the sample value.

PCM: Pulse-Code Modulation. A modulation technique that encodes the amplitude of a sampled signal in a digital format.

PDM: Pulse Duration Modulation - another name for PWM.

Phase-Splitter: A mechanism to provide complementary signals to a next circuit stage. E.g., a base-driven transistor with both an emitter and a collector load with outputs derived from both.

Phonon: An acoustic lattice vibration.

Photon: The quantized particle embodiment of light, with energy equal to the product of Planck's constant and the light frequency.

Phototransistor: A transistor packaged so that its base region (for bipolar) can be illuminated by incoming signal light. The incident photons induce hole-electron pairs that behave much as would electrically injected base charge. Unipolar phototransistors also exist. Both device types tend to be slower than diode detectors but are capable of power gain.

Pigtail: (In the context of this text.) A length of fiber coupled to a device (e.g., a light transmitter or receiver).

Pinning: Clamping.

Poisson Distribution: A distribution whose mean is equal to its variance. Corresponds to a random arrival regime.

Positron: A particle with the mass and charge magnitude of an electron but of opposite charge polarity: positive.

Positron-decay: A nuclear decay process that emits a positron and a neutrino.

Postamplifier: An amplifier following a preamplifier. The preamplifier functions to render the signal more robust relative to noise, and the postamplifier merely increases the signal's power level.

Poynting Vector: A vector whose magnitude and direction are given by the cross product of the **E** and **H** fields at a given point. It yields the direction of propagation and magnitude of the power of a propagating electromagnetic wave.

PPM: Pulse-Position Modulation - A modulation technique which represents the amplitude of a sample with a pulse whose temporal position relative to a nominal point in time is proportional to the sample amplitude.

Preamplifier: The first amplifier following a transducer whose output is so feeble as to require special treatment; e.g., boosting of the electrical signal close to the point of origin to prevent excessive noise ingress while the signal is at a low level.

Preform: The solid cylinder of processed glass from which fiber may be pulled.

Pulse-Analog Modulation: Modulation using one of the pulse-oriented script techniques: PDM, PPM, or PFM.

PWM: Pulse Width Modulation - A modulation technique which represents the amplitude of a sample with a pulse whose width is proportional to the amplitude of the sample.

Quantizing Noise: The noise resulting from the imperfect representation of a continuously ·varying analog signal sample by a finite-length digital string. Characteristic of PCM and its variations.

Quantum Efficiency: Efficiency at the particle level in conversion to or from the electrical domain from or to the optical domain.

R: Responsivity (see below).

Rad: A radiation dosage level corresponding to the absorption of ionizing radiation in any medium of 100 ergs/g.

Raman Scattering: Scattering of light in which a shift in wavelength from that of the nominally monochromatic propagating wave occurs. The amount of shift is a function of the scattering particles and wavelengths. Significant only at very high power densities.

Rayleigh Scattering: Scattering induced by random fluctuations at the molecular level; inversely proportional to the fourth power of the wavelength.

Reflection Coefficient: A generally complex scalar indicating the amplitude and phase of a reflected signal relative to the incident signal.

Responsivity (R): The ratio between photodetector photocurrent and incident optical power. Units of amperes per watt.

Rho (ρ): Charge density. Units of coulombs per cubic meter.

RZ: Return to Zero. A binary signal format where the signal returns to "zero" between successive "one" signals.

SECAM: A television standard that has been adopted, for example, by France and the Socialist countries. The color subcarrier is frequency modulated, alternately, by the color difference signals.

Sigma (σ): Conductivity. Units of siemens per meter.

Silica: Silicon dioxide.

Single-event upset: Loss of a bit in a digital stream due to the effects of radiation.

Single-Mode Fiber: Fiber whose core is of such dimensions relative to the wavelength of the transmitted light that only one mode (the HE_{11}) can be supported. Characterized by a V number of 2.405 or less.

Skew Ray; A light ray that does not intersect the axis of the fiber.

Slope Overload: The phenomenon which occurs when the rate of change in amplitude of a modulating signal overreaches the ability of simple Delta-Modulation to convey the information.

SMPTE: Society of Motion Picture and Television Engineers.

Snell's Law: Law of refraction:

$$n_1 \sin \phi_1 = n_2 \sin \phi_2 ,$$

where the n's and ϕ's are the indices of refraction and angles to the normal on the two sides of the interface.

Sniper-Scope: An image-intensifying device used for seeing in low ambient light conditions; applicable for detecting infrared emanations.

Soot: In the context of optical fiber manufacture, the glassy residue of the flame-oxidation process which is deposited upon the preform-in-the-making.

Space: The presence of a signal intended to represent a logical 0.

Speckle Pattern: The pattern of light intensity in a plane orthogonal to the direction of transmission due to the finite mode set supported by illumination with a coherent source.

Stagger-Tuning: A technique which tunes successive stages of an amplifier at differing frequencies to yield a relatively flat, wide-band gain characteristic.

Step-Index Fiber: Fiber in which an abrupt step in refractive index occurs at the core-cladding interface.

Stimulated Emission: Emission of light as the result of a photon impinging upon an electron in an elevated energy state and stimulating it to revert to a lower energy state, causing a photon with a corresponding energy, direction of propagation, and phase, to join the incident photon.

Strobe: The act of selecting one or more temporal segments of a signal.

Summing Circuit: A circuit which combines its inputs to produce an output amplitude which is the arithmetic sum of the input amplitudes.

Superlinear: Increasing at a greater than linear rate. A possible characteristic of the output light power as a function of drive current for an edge-emitting diode.

Superluminescent: Superlinear behavior of a light-emitting diode.

TASI: Time Assignment Speech Interpolation. A technique which dynamically steals channels away from non-articulating speakers in a multiplexed system and provides them to those who are articulating. Because (usually) one party to a conversation is listening while the other is speaking, this technique should at least double the channels; in practice, it roughly trebles them.

Telidon: A Canadian version of Videotex.

TE Mode: A propagation mode with the property of having no axial electric field component.

Threshold Current: The current level below which spontaneous emission occurs in an ILD and above which stimulated emission occurs. This level is sensitive to both age and temperature.

Time-Domain Reflectometer: A device which can launch a brief pulse of light into a fiber and display the backscattered and reflected result. Useful for determining characteristics and detecting anomalies in fiber.

TM Mode: A propagation mode with the property of having no axial magnetic field component.

Totem Pole: In an active circuit, a configuration comprising an active device to pull up an output, coupled with an active device to pull it down. Typified by the output stage of a T^2L logic gate: an emitter-follower structure to pull up, and a common emitter structure to pull down.

Transducer: A device capable of converting one form of power into another.

Transimpedance Amplifier: A feedback amplifier structure typically used as a preamplifier which transforms an input current signal at high impedance into an output voltage signal at low impedance. These structures have a wide bandwidth, a wide dynamic range, and often do not require equalization.

Trapping State: An energy state intermediate in level between the valence band and conduction band which is capable of trapping an electron descending from the conduction band.

U: The product of the transverse propagation constant in the core and the core radius (dimensionless).

V: Normalized frequency $[= ka(NA)]$ (dimensionless).

VAD: Vapor Axial Deposition. A fiber manufacturing process that deposits doped silica via a vapor-phase oxidation process at the lower extremity of an upward-moving glass rod.

Valence Band: The energy band in a semiconductor representing the ground state relative to electron excitation.

Videotex: An interactive service that displays user-requested, computer-derived information on a Television screen. The signals are conveyed over conventional telephone lines.

Videotext: A technique for broadcasting information using the horizontal lines blanked during the vertical interval in a TV frame. The information is primarily in character form, though low resolution figures may be conveyed.

W: The product of the transverse decay constant in the cladding and the core radius (dimensionless).

Wavelength (λ): The distance between periodic maxima (or minima) in transmitted waves. Typically expressed in terms of its free-space value.

Wentzel-Kramers-Brillouin Method: An approximation method useful in solving wave equations.

WKB: Wentzel-Kramers-Brillouin Method. (See above.)

X-ray: Radiation of wavelength smaller than visible light but greater than gamma rays. Usually produced by the impact of high energy electrons upon a metallic target.

BIBLIOGRAPHY

Agrawal, G. P. and Dutta, N. K., *Long-Wavelength Semiconductor Lasers*, Van Nostrand Reinhold, 1986.

Akamatsu, T., "Continuous Fabrication of a Phosphate Glass Fiber," *Journal of Lightwave Technology*, vol. LT-1, no. 4, December, 1983, pp. 580-584.

Albanese, A. and Lenzing, H. F., "IF Lightwave Entrance Links for Satellite Earth Stations," *IEEE International Communications,* 1979, p. 1.7.1.

Albanese, A., "An Automatic Bias Control (ABC) Circuit for Injection Lasers," *Bell System Technical Journal*, vol. 57, 1978, pp. 1533-1544.

Alferness, R. C., Joyner, C. H., and Buhl, L. L., "High Speed Traveling-Wave Directional Coupler Modulator for $\lambda = 1.32$ μms," *1983 Optical Fiber Communication Meeting, Optical Society of America,* p. 20.

Anderson, C. D., Gleason, R. F., Hutchison, P. T., and Runge, P. K., "An Undersea Communication System Using Fiber-Guide Cables," *Proceedings of the IEEE*, vol. 68, no. 10, 1980, pp. 1299-1303.

Anderson, W. T., et al., "Mode-Field Diameter Measurements for Single-Mode Fibers with Non-Gaussian Field Profiles," *Journal of Lightwave Technology*, vol. LT-5, no. 2, February, 1987, pp. 211-217.

Ando et al., "Characteristics of Germanium Avalanche Photodiodes in the Wavelength Region of 1-1.6 μms," *IEEE Journal of Quantum Electronics,* November, 1978, pp. 804-809.

Aoki, F. and Nabeshima, "Optical Fiber Communications for Electric Power Companies in Japan," *Proceedings of the IEEE*, vol. 68, no. 10, 1980, pp. 1280-1285.

Arnold, G. and Krumpholz, O., "Coupling of Monomode Fibers to Edge-Emitting Diodes," *Optical Fiber Communications/Optical Fiber Sensors,* February, 1983, pp. 48-49.

Asatani, K., Nosu, K., Matsumoto, T., and Yanagiomota, K., "A Field Trial of Fiber Optic Subscriber Loops in Yokosuka," *ICC '81 Conference,* vol. 3, Paper 48.1.1-5.

Atal, B. S. and Hofacker, R. Q., Jr., "The Telephone Voice of the Future," *AT&T Bell Laboratories Record*, July, 1985, pp. 4-10.

Baack, C., et al., "Analogue Optical Transmisssion of 26 TV Channels," *Electronics Letters*, May 10, 1979, vol. 15, no. 10, pp. 300-301.

Baden Fuller, A. J., *An Introduction to Microwave Theory and Techniques,* Pergamon Press, 1979, pp 135-39.

Barnes, C. E., *Radiation Effects in Optoelectronic Devices*, Sandia Report, SAND76-0726, 1976, pp. 6 & 126.

Barnowski, M. K., "Fiber Systems in the Military Environment," *Proceedings of the IEEE*, vol. 68, no. 10, 1980, pp. 1315-1320.

Barnowski, M. K., *Fundamentals of Optical Fiber Communication,* Academic Press, Inc., 1972, p. 92.

Bartee, T. C., *Digital Communications*, Sams, 1986, pp. 1-48.

Basch, E. E., Beaudette, R. A., and Larnes, H. A., "Optical Transmission for Interoffice Trunks," *Transactions of the IEEE Communications Society*, vol. 26, no. 7, 1978, pp. 1007-1014.

Belcher and Marshall, "Use of Fiber Optics in Digital Automatic Flight Control Systems," *IEEE Transactions on Aerospace Engineering Science,* September, 1975, pp. 841-50.

Bell, A. G., "Selenium and the Photophone," *The Electrician*, vol. 5, 1880, p. 214ff.

Bell, T. E., *IEEE Spectrum,* January, 1985, pp. 53-57.

Bendow, B. and Mitra, S. S., *Fiber Optics,* Plenum press, 1979, pp. 15-17.

Berriman, D. W, "A Lens or Light Guide Using Convectively Distorted Thermal Gradients in Gases," *Bell System Technical Journal*, vol. 43, no. 4, July, 1964, pp. 1469-1475; and "Growth of Oscillations of a Ray About the Irregularly Wavy Axis of a Lens Light Guide," vol. 44, no. 9, November, 1965, pp. 2117-2132.

Bhagavatula, V. A., Spatz, M. S., Love, W. F., and Keck, D. B., Segmented-Core Single-Mode Fibres with Low-Loss and Low Dispersion," *Electronics Letters*, vol. 19, 1983, pp. 317-318.

Bly, D., "Fiber is Cost-Effective Now," *Telephone Engineer and Management,* vol. 89, no. 1, January, 1985, p. 52.

Boggs, L. M. and Buckler, M. J., "Testing Lightguide Fiber," *Western Electric Engineer*, vol. XXIV, no. 1, Winter, 1980.

Bohn, P. P. et al., "Bringing Lightwave Technology to the Loop," *Bell Laboratories Record,* April, 1983, pp. 6-10.

Born, M. and Wolf, E., *Principles of Optics,* 4th ed., Pergamon Press, 1970, pp. 110ff.

Bowmand, R. L., and Lane, J. L., "Fiber Optics - An Exploding Industry," *Telephone Engineer and Management,* February, 1985, p. 115.

Briley, B. E., *Introduction to Telephone Switching,* Addison-Wesley, 1983, p. 2.

Bruce, R. V., *Alexander Graham Bell and the Conquest of Solitude,* London: Gollancz, 1973.

Buetikofer, J. and Stettler, U., "Pilot Network for Wideband Communications at Marsens, Switzerland," *Tech Mitt PTT,* vol. 61, no. 12, 1983, pp. 414-420.

Burrus, C. A. and Dawson, R. W., "Small-Area High-Current Density GaAs Electroluminescent Diodes and a Method of Operation for Improved Degradation Characteristics," *Applied Physics Letters,* vol. 17, 1970, pp. 97-98.

Carlson, A. B., *Communication Systems,* 2nd Ed., McGraw-Hill, 1975, pp. 236-237.

Carroll, J. M., *The Story of the LASER,* E. P. Dutton & Co., 1964, p. 123.

Caton, W. M., "Fiber Optics Systems in Adverse Environments, II, "*Proceedings of the International Society for Optical Engineering,* August, 1981.

Chang, K. Y., "Fiberguide Systems in the Subscriber Loop," *Proceedings of the IEEE,* vol. 68, no. 10, 1980, pp. 1291-1299.

Chang, K. Y. and Hara, E., "Fiber-Optic Broad-Band Integrated Distribution - Elie and Beyond," *IEEE Journal on Selected Areas in Communications,* vol. SAC-1, April, 1983, pp. 439-44.

Cheng, D. K., *Field and Wave Electromagnetics,* Addison-Wesley, 1983, pp. 326ff.

Cherin, A. H., *An Introduction to Optical Fibers,* McGraw-Hill, 1983, pp. 52ff.

Chida, K., Hanawa, F., and Nakahara, M., "Fabrication of OH-Free Multimode Fiber by Vapor Phase Axial Deposition," *IEEE Journal on Quantum Electronics,* November, 1982, pp. 1883-89.

Clarricoats, P. J. B., ed., *IEE Reprint Series 1: Optical Fibre Waveguides,* Peter Peregrinus, Ltd., 1975, Cover Page.

Cohen, L. G., Marcuse, D, and Mammel, W. L., "Radiating Leaky-Mode Losses in Single-Mode Lightguides with Depressed Index Claddings," *IEEE Journal of Quantum Electronics*, vol. QE-18, no. 10, October, 1982, pp. 1467-1472.

Cunningham, D. J. and Lymphany, S. S., "Ethernet: Fiber Optic Design Issues and Applications Experience," *FOC/LAN 84*, September, 1984, pp. 101-108.

Dakin, J. P., "Optical Fiber Sensors - Principles and Applications," *Proceedings of the SPIE - The International Society for Optical Engineering*, vol. 374, 1983, pp. 172-182.

Daly, J. C., "Fiber Optic Intermodulation Distortion," *IEEE Transactions on Communications,* vol. COM-30, no. 8, August, 1982.

Davies, K., "Ionospheric Radio Propagation," *Monograph 80*, National Bureau of Standards, Washington, D.C., April, 1965.

Dezelsky, F. T., Sprow, R. B., and Topolski, F. J., "Lightguide Packaging," *Western Electric Engineer*, Winter, 1980, pp. 81-85.

Dupuis, R. D., Moudy, L. A., and Dapkus, P. D., "Preparation and Properties of Ga(1-x)Al(x)As-GaAs Heterojunctions Grown by Metal-Organic Chemical Vapor Deposition," *Gallium Arsenide and Related Compounds 1978*, Inst. Phys. Conf. Ser., vol. 45, pp. 1-9.

Epworth, R. E., "The Phenomenon of Modal Noise In Analogue and Digital Optical Fiber Systems," *4th European Conference on Optical Communications,* Genoa, 1978.

Ettenberg and Kressel, H., *Journal of Applied Physics*, vol. 47, 1976, p. 1538.

Ettenberg, M., Kressel, H., and Wittke, J. P., "Very High Radiance Edge-Emitting LED," *IEEE Journal on Quantum Electronics,* June, 1976, p. 360.

Fitchew, K. D., "Technology Requirements for Optical Fiber Submarine Systems," *IEEE Journal on Selected Areas in Communications,* vol. SAC-1, no. 3, April, 1983, p. 447.

Flint, G. W., "Analysis and Optimization of Laser Ranging Techniques," *IEEE Military Electronics,* January, 1964, p. 22.

Fox, J. R., Fordham, D. L., Wood, R., and Ahern, D. J., "Initial Experience with the Milton Keynes Optical Fiber Cable TV," *IEEE Transactions on Communications,* vol. COM-30, no. 9, September, pp. 2155-2162.

Freyhardt, H. C., Ed., *Crystals: Growth, Properties, and Applications*, Springer-Verlag, 1980, vol. 3, pp. 73-162.

Friebele, E. J., Schultz, P. C., Gingerich, M. E. and Hayden, L. M., "Effect of B, P, and OH on the Radiation Response of GE-Doped Silica-Core Fiber-Optic Waveguides," *Digest, Topical Meeting on Optical Fiber Communication*, March 1979, Washington, DC., p. 36.

Friebele, E. J., et al., "Radiation Damage in Single-Mode Optical Fiber Waveguides," *Digest of Conference on Optical Fiber Communications*, Washington, D.C., 1982, pp. 1-9.

Fujii, Y., et al., "Low-Loss 4X4 Optical Matrix Switch for Fibre-Optic Communication," *Progress in Optical Communication*, IEE Reprint Series 3, pp. 184-85.

Gallawa, R. L., "U.S. and Canada: Initial Results Reported," *IEEE Spectrum*, vol. 16, no. 10, October, 1979, pp. 72-73.

Gallegher, R. T., "Cabled City Forms French Prototype," *Electronics*, June 14, 1984, pp. 88-92.

Gardner, W. B., "Microbending Loss in Optical Fibers," *Bell System Technical Journal*, vol. 54, no. 2, February, 1975, pp. 452-465.

Ginsburg, C. P., "Report of the SMPTE Digital Television Study Group," *SMPTE Technical Conference*, Detroit, 1976.

Glodis, P. F., Anderson, W. T., and Nobles, J. S., "Control of Zero Chromatic Dispersion Wavelength in Fluorine-Doped Single-Mode Optical Fibers," *1983 Optical Fiber Communication Meeting, Optical Society of America*, IEEE #83CH1850-7, p. 12.

Glodis, P. F. et al., "Bending Loss Resistance in Single-Mode Fiber," *OFC/IOOC Technical Digest*, TUA3, p. 41.

Gloge, D., "Weakly Guiding Fibers," *Applied Optics*, vol. 10., 1971, pp. 2252-2258.

Godfrey, L. A. and Garside, B. K., "Optimal Design of Ultrafast Photodetectors," *Conference on Quantum Electron Lasers and Electronics*, June, 1981, p. 66,

Goel, J. E., "An Optical Repeater with High-Impedance Input Amplifier, *Bell System Technical Journal*, vol. 53, no. 4, April 1974, p. 640.

Gordan, J. P., Zeiger, H. J., and Townes, C. H., "Molecular Microwave Oscillator and New Hyperfine Structure in the Microwave Spectrum of NH3, "*Physical Review*, vol. 95, 1954, p. 282.

Gosch, J., "Silicon Carbide Ends Long Quest for Blue Light-Emitting Diode," *Electronics Week,* October 8, 1984, p. 24.

Gover, J. E. and Srour, J. R., *Basic Radiation Effects in Nuclear Power Electronics Technology*, Sandia Report, SAND85-0776, May, 1985.

Gregg, W. D., *Analog and Digital Communication*, Wiley, 1977, p. 76.

Greiling, P. T., "The Future Impact of GaAs Digital IC's," *IEEE Journal on Special Topics*, vol. SAC-3, no. 2, March, 1985, pp. 384-393.

Haibara, T., "Fully Automatic Optical Fibre Arc-Fusion Splice Machine," *Electronics Letters,* vol. 20, no. 25-26, December, 6, 1984, pp. 1065-66.

Hall, R. N., Genner, G. E., Kingsley, J. D., Soltys, T. J., and Carlson, R. O., "Coherent Light Emission from GaAs Junctions," *Physics Review Letters*, vol. 9, 1962, p. 366.

Hawks, D., *Pioneers of Wireless,* Methner & Co. Ltd., pp 60-86.

Hayashi, I., Panish, M. B., Foy, P. W. and Sumelay, S., "Junction Lasers Which Operate Continuously at Room Temperature," *Applied Physics Letters,* vol. 17, no. 3, 1970, pp. 109-111.

Hecht, J., "Victorian Experiments and Optical Communications," *IEEE Spectrum,* vol. 22, no. 2, February, 1985, p. 69.

Henry, P. S., "Lightwave Primer," *IEEE Journal on Quantum Electronics*, vol. QE-21, 1985, pp. 1862-1879.

Henschied, A., Barney, P. J., "CATV Applications of Feedforward Techniques," *IEEE Transactions on Cable Television,* vol. CATV-5, no. 2, April, 1980, pp. 80-85.

Hirai, M., Kawase, W., Kobayashi, H. and Katsuyama, Y., "Optical Fiber Cables for Local Area Network," *IEEE International Conference on Communications (ICC '83)*, June 1983, pp. 707-712.

Holonyak, N., Jr. and Bevelaqua, S. F., "Coherent (Visible) Light Emission from $Ga(As[1-x]P[x])$ Junction," *Applied Physics Letters*, vol. 1, 1962, p. 82.

Horiguchi, T., "Optical Time Domain Reflectometer for Single-Mode Fibers," *Transactions of the Institute of Electronics and Communication Engineers, Japanese Sect. E.,* vol. 67, no. 9, September, 1984, pp. 509-515.

Hoshikawa, M., "Optical Fiber Splicing," *Japanese Annual Review of Electronic Computers and Telecommunications,* vol. 5, 1983, pp. 209-218.

Inoue, T., Koizumi, K., and Ikeda, Y., "Low-Loss Light-Focusing Fibres Manufactured by a Continuous Process," *Proceedings of the IEE*, vol. 123, no. 6, June, 1976, pp. 577-580.

Ito, T., and Nakagawa, K., "Transmission Experiments in the 1.2-1.6 μm Wavelength Region Using Graded-Index Optical Fiber Cables," *Fiber and Integrated Optics*, vol. 3, no. 1, 1980, p. 7.

Iwashita, K., et al., "Linewidth Requirement Evaluation and 290 km Transmission Experiment for Optical CPFSK Differential Detection," *Electronics Letters*, vol. 22, July, 1986, pp. 791-792.

Jacobs, I. and Stauffer, J. R., "FT3 - A Metropolitan Trunk Lightwave System," *Proceedings of the IEEE*, vol. 68, no. 10, 1980, pp. 1286-90.

Jeunhomme, L. B., *Single-Mode Fiber Optics, Principles and Applications*, Marcel Dekker, Inc. 1983, p. 47.

Kabziw, J., "BIGFON: Preparation for the Use of Optical Fiber Technology in the Local Network for the Deutsche Bundespost," *IEEE Journal on Selected Areas in Communications,* vol. SAC-1, no. 3, April, 1983, pp. 436-439.

Kalish, D. and Cohen, L. G., "Single-Mode Fiber: From Research and Development to Manufacturing," *AT&T Technical Journal*, vol. 66, issue 1, January/February, 1987, pp. 19-32.

Kaminow, I. P., Eisenstein, G., Stulz, L. and Denki, A., "Lateral Confinement in InGaAsP Super-Luminescent Diode at 1.3 Microns," *IEEE Journal on Quantum Electronics*, vol. QE19, no. 1, January, 1983, pp. 78-82.

Kao, K. C. and Hockham, G. A., "Dielectric Fibre Surface Waveguides for Optical Frequencies," *Proceedings of the IEEE*, vol. 113, July, 1966, pp. 1151-1158.

Kapron, F. P., "Fiber-Optic System Tradeoffs," *IEEE Spectrum,* March, 1985, p. 70.

Kapron, F. P., Keck, D. B., and Maurer, R. D., "Radiation Losses in Glass Optical Waveguides," *Applied Physical Letters*, vol. 17, 1970, pp. 423-425.

Kapron, F. P., *Spectrum,* March, 1985, p. 70.

Kawahata, M., "Fiber Optics Application to Full Two-Way CATV System - HI-OVIS," *National Telecommunications Conference Record*, 1977, p. 14:4.

Keck, D. B. and Bhagavatula, "Single Mode Fiber Design," *Proceedings of the Sixth Topical Meeting on Optical Fiber Communications*, New Orleans, February, 1984, Paper MF1.

Keck, D. B., Schultz, P. C., and Zimar, F., *U. S. Patent No. 3,737,292.*

Klein, M. V., *Optics*, John Wiley and Sons, 1970, p. 534.

Kobayashi, S. and Kimura, T., "Semiconductor Optical Amplifiers," *IEEE Spectrum*, May, 1984, pp. 26-33.

Kock, W. E., *Engineering Applications of Lasers and Holography*, Plenum, 1975, p. 331.

Kogelnik, H. and Shank, C., "Coupled-Wave Theory of Distributed Feedback Lasers," *Journal of Applied Physics*, vol. 43, 1972, p. 2327.

Koizumi, K., Ikeda, Y., Kitano, I., Furukawa, M., and Sumimoto, T., "New Light-Focusing Fibers Made by a Continuous Process, *Applied Optics*, vol. 13, no. 2, February, 1974, pp. 255-260.

Kressel, H., *Topics in Applied Physics, Volume 39, Semiconductor Devices for Optical Communication*, Springer Verlag, 1980, pp. 54-55, 125-26, 139, 260-61, 266ff.

Kroemer, H., "A Proposed Class of Heterojunction Injection Lasers," *Proceedings IEEE*, vol. 51, 1963, p. 1782.

Kroemer, H., "Heterostructure Bipolar Transistors and Integrated Circuits," *Proceedings of the IEEE, Special Issue on Very Fast Solid-State Technology*, vol. 70, no. 1, January, 1982, pp. 13-25.

Lanahan, T. A., "Calculation of Modes in an Optical Fiber Using the Finite Element Method and EISPACK," *Bell System Technical Journal*, vol. 62, no. 9, Part 1, November, 1983, p. 2664.

Lee, T. P., Burrus, C. A., Dentai, A. G., "Dual Wavelength Surface Emitting InGaAs L.E.D.s," *Electronics Letters*, October 23, 1980, vol. 16, no. 22, pp. 845-846.

Lemaire, P. J. and Tomita, A, "Behavior of Single Mode Fibers Exposed to Hydrogen," *Proceedings of the Tenth European Conference on Optical Communication*, Stuttgart, September 3-6, 1984.

Li, T., "Advances in Optical Fiber Communications: An Historical Perspective," *IEEE Journal on Selected Areas of Communication*, vol. SAC-1, no. 3, April, 1983, pp. 356-72.

Li, T., "Advances in Lightwave Systems Research," *AT&T Technical Journal*, January/February, 1987, vol. 66, Issue 1, pp. 5ff.

Lines, P. W. and Millington, D., "Optical Fiber Systems for the British Telecom Trunk and Junction Networks," *National Telecommunications*, IEEE, 1980, p. 46.4.3.

Linke, R., Jasper, B., Ko, J., Kaminow, I. and Vodhanel, R., *Proceedings of the 4th International Conference on Integrated Optics and Optical-Fiber Communications*, Tokyo, 1983.

Maclean, D. J. H., *Broadband Feedback Amplifiers*, Research Studies Press, 1982, p. 181.

Maiman, T. H., "Stimulated Optical Radiation in Ruby Masers," *Nature*, vol. 187, 1958, p 493.

Marchesi, C. et al., "Analogue Transmission of Video Signals and Applications to Telemedicine," *CSELT Technical Report*, vol. XII, no. 2, April 1984, pp. 189-191.

Mattern, P. L. et al, "Effects of Radiation on Absorption and Luminescence of Fiber Optic Waveguides and Materials," *IEEE Transactions on Nuclear Science*, vol. NS-21, no. 6, 1974, pp. 81-85.

McChesney, J. B., "Materials and Processes for Preform Fabrication - Modified Chemical Vapor Deposition and Plasma Chemical Vapor Deposition," *Proceedings of the IEEE*, vol. 68, no. 10, October, 1980, pp. 1182-1183.

Mesiya, M. F., Miller, G. E., Pinnowe, D. A., "Mini-Hub Addressable Distribution System for Hi-Rise Application," *Technical Papers: Cable'82, NCTA 31st Annual Convention & Exposition*, May, 1982, pp. 37-42.

Meyer, R. G., Eschenbach, R., Edgerly, W. M., Jr., "A Wide-Band Feedforward Amplifier," *IEEE Journal on Solid-State Circuits*, vol. SC-9, no. 6, December, 1974, pp. 422-428.

Michalopoulos, D., "Fiber Telecommunications System Reduces Need for Signal Regeneration," *Computer*, March, 1982, p. 96.

Midwinter, J, E., "Optical Fibre Communications, Present and Future," *The Clifford Paterson Lecture, Proceedings of the Royal Society: London*, A 392, 1984, pp. 247-277.

Midwinter, J. E., *Optical Fibers for Transmission*, John Wiley & Sons, Inc., 1979, p. 188.

Mies, E. W. and Soto, L., "Characterization of the Radiation Sensitivity of Single-Mode Optical Fibers," *11th European Conference on Optical Communication*, Venice, Italy, 1985.

Miki, T. et al., *Technical Digest of the International Conference on Integrated Optics and Optical Fiber Communications*, Japan, 1977.

Miller, C. M., "Loose Tube Splices for Optical Fibers," *Bell System Technical Journal*, vol. 54, no. 7, September, 1975, pp. 1215-1225.

Miller, C. M., *Optical Fiber Splices and Connectors*, Marcel Decker, Inc., 1986.

Miller, C. M. and DeVeau, G. F., "Simple High-Performance Mechanical Splice for Single-Mode Fibers," *Technical Digest: Conference on Optical Fiber Communication*, San Diego, February 11, 1985, Paper MI2, p. 21.

Miller, S. E. and Chynoweth, A. G., *Optical Fiber Telecommunications*, Academic Press, Inc., 1979, p. 21.

Miller, S. E., "Present Thrust of Optical-Fiber Telecommunications Research," *IEEE Journal of Lightwave Technology: An Individual Perspective,* vol. LT-2, no. 4, August, 1984, pp. 494-95.

Miyamoto, M. et al., "Effects of Hydrogen on Long-Term Reliability of Optical Fiber Cable," *Conference on Optical Fiber Communications,* San Diego, February, 1985, p. 46.

Monclavo, A. and Tosco, F., "European Field Trials and Early Applications in Telephony," *IEEE Journal on Selected Areas in Communication*, vol. SAC-1, April, 1983, pp. 398-403.

Moriyama, T, Fukuda, O., Sanada, K., Inada, K., Edahiro, H., and Chida, K., "Ultimately Low OH Content VAD Optical Fibers," *Electronics Letters*, vol. 16, August, 1980, pp. 699-700.

Morse, P. M. and Feshback, H., *Methods of Theoretical Physics*, McGraw-Hill, 1953, p. 1092.

Myashita, T., Miya, T, and Nakahara. M., "An Ultimate Low Loss Single Mode Fiber at 1.55 μms, *Optical Fiber Communications Conference Proceedings*, Washington, D.C., March 6-8, 1979.

Nantz, T. D., and Shenk, W. J., "Lightguide Applications in the Loop," *AT&T Technical Journal*, vol. 66, issue 1, January/February, 1987, pp. 108-118.

Nomura, H., "Very Fast LEDs Offer Response for Optical Local Area Networks," *Journal of Electronic Engineering*, vol. 22, no. 217, January, 1985, pp. 34-38.

O'Connor, P. B. et al., "Large Core High NA Fibers for Data Link Applications," *Progress in Optical Communication*, IEE Reprint Series 3, p. 196.

Ogawa, K., Lee, T. P., Burrus, C. A., Campbell, J. C., Dentai, A. G., "Wavelength Division Multiplexing Experiment Employing Dual-Wavelength LEDs and Photodetectors," *Electronics Letters*, October 29, 1981, vol. 17, no. 22, pp. 857-858.

Okoshi, T., *Optical Fibers*, Academic Press, 1982, p. 59.

Okoshi, T., "Recent Advances in Coherent Optical Fiber Communication Systems," *Journal of Lightwave Technology*, vol. LT-5, no. 1, January, 1987, pp. 44-52.

Olshansky, R., and Keck, D. B., "Pulse Broadening in Graded-Index Optical Fibers," *Applied Optics*, vol. 15, February, 1976, pp. 483-491.

Olsson, N. A., "Single Frequency Lasers and 20 Gb/s Optical Transmission Experiments," *National Science Foundation Workshop and Grantee User's Meeting on Optical Communication Systems*, Ithica, New York, June, 1985.

Panter, P. F., *Modulation, Noise, and Spectral Analysis*, McGraw-Hill Book Company, 1965, pp. 662-673.

Payne, D. N. and Gambling, W. A., New Low-Loss Liquid-Core Fibre Waveguide," *Electronics Letters*, vol. 8, 1972, pp. 374-376.

Peebles, P. Z., *Communication System Principles*, Addison-Wesley, 1976, p. 393ff.

Personick, S. D., "Photon Probe - An Optical-Fiber Time-Domain Reflectometer," *Bell System Technical Journal*, vol. 56, no. 3, March, 1977, pp. 355-366.

Personick, S. D., *Optical Fiber Transmission Systems*, Plenum Press, 1981, p. 157.

Peterson, G. E., "Electrical Transmission Lines as Models for Soliton Propagation in Materials: Elementary Aspects of Video Solitons," *AT&T Bell Laboratories Technical Journal*, no. 6. Part 1, July-August, 1984, pp. 901-919.

Peterson, R. C. and Sliney, D. H., "Toward the Development of Laser Safety Standards for Fiber-Optic Communications Systems, *Applied Optics*, vol. 25, no. 7, April 1, 1986, pp. 1038-1047.

Piasetsky, J. and Murphy, R., "DCM Systems to Expand TAT-8 Circuit Capacity," *Telephony*, August 4, 1986, pp. 70-74.

Quick, W. H., James, K. A., and Coken, J. E., "Fiber Optics Sensing Techniques," *First International Conference on Optical Fiber Sensors*, April, 1983, pp. 6-8.

Rashleigh, S. C. and Stolen, R. H., "Status of Polarization-Preserving Fibers," *CLEO '84 Conference Proceedings*.

Refi, J. J., "Fiber Bandwidth and its Relation to System Design," *FOC/LAN 86*, Orlando, Florida, October, 1986, pp. 251-257.

Reitz, P. R., "Characterization of Concatenated Multimode Optical Fibers with Time Domain Measurement Techniques," *Proceedings of Conference on Precision E-M Measurements*, Boulder, Colorado, 1982, p. L-14.

Runge, P. K. and Trischitta, P. R., "The SL Undersea Lightwave System," *IEEE Journal on Selected Areas in Communication,* vol. SAC-2, no. 6, November, 1984, pp. 784-793.

Sakurai, K. and Asatani, K., "A Review of Broad-Band Fiber System Activity in Japan," *IEEE Journal on Selected Areas in Communication*, vol. SAC-1, no. 3, April 1983, pp 428-435.

Sanferrare, R. J., "Terrestrial Lightwave Systems," *AT&T Technical Journal*, January/February, 1987, vol. 66, issue 1, p. 99.

Sanger, C. C. and Williams, D. L., *Fibre Optics 1983,* April 19-21, 1983, p. 141.

Schepers, C., "A Digital CTV Chassis Concept," *1983 IEEE International Conference on Consumer Electronics: Digest*, pp. 88-9.

Schmidt, R. V. and Alferness, R. C., "Directional Couplers, Switches, Modulators, and Filters Using Alternating Delta-Beta Techniques," *IEEE Transactions on Circuits and Systems,* vol. CAS-26, no. 12, December, 1979, pp. 1099-1108.

Schwartz, M., *Information Transmission, Modulation, and Noise,* McGraw-Hill, 1980, pp. 147-49.

Scott-Russell, J., "Report on Waves," *Proceedings of the Royal Society of Edinburgh,* 1844, pp. 319-320.

Sears, F. M. and Simpson, J. R., "Polarization Quality of High-Birefringence Single-Mode Fibers," *AT&T Bell Laboratories Technical Journal*, February, 1984, vol. 63, no. 2, pp. 365-371.

Senior, J. M., *Optical Fiber Communications Principles and Practices*, Prentice-Hall, 1985, p. 21.

Simon, J. C., "Semiconductor Laser Amplifier for Single Mode Optical Fiber Communications," *Journal of Optical Communications*, April, 1983, pp. 51-62.

Schelkunoff, S. A., *Electromagnetic Waves*, D. Van Nostrand Company, 1943.

Slepian, D., "On Delta Modulation," *Bell System Technical Journal*, vol. 51, no. 10, December, 1972, p. 2101.

Smith, R. G., "Optical Power Handling Capability of Low Loss Optical Fiber as Determined by Stimulated Raman and Brillouin Scattering," *Applied Optics*, vol. 11, November, 1972 pp. 2489-2494.

Snyder, A. W., "Asymptotic Expressions for Eigenfunctions and Eigenvalues of a Dielectric or Optical Waveguide," *Transactions, IEEE Microwave Theory Tech.*, MTT-17, 1969, pp. 1130-1138.

Stanley, W. D., *Electronic Communications Systems*, Reston, 1982, pp. 197-98.

Stremler, F., *Introduction to Communication Systems*, Addison-Wesley Publishing Co., p. 365.

Sze, S. M., *Physics of Semiconductor Devices, 2nd Edition*, John Wiley and Sons, 1981, pp. 689-700.

Tada, K., and Mirose, K., "A New Light Modulator using Perturbations of Synchronism Between Two Coupled Guides," *Applied Physics Letters*, vol. 25, November, 1974, pp. 561-562.

Tamir, T., *Topics in Applied Physics, Vol. 7: Integrated Optics*, Springer-Verlag, 2nd Edition, 1979, p. 16ff.

Tanifuji, T. and Kato, Y., "Realization of a Low-Loss Splice for Single-Mode Fibers in the Field Using an Automatic ARC-Fusion Splicing Machine," *1983 Optical Fiber Communication Meeting, Optical Society of America*, IEEE # 83CH1858-7, p. 14.

Tasker, G. W. and French, W. G., "Low-Loss Optical Waveguides with Pure Fused Silicon Dioxide Cores," Proceedings of the IEEE, vol. 62, 1974, p. 1281.

Taylor, A. S., "Characterization of Cable TV Networks as the Transmission Media for Data," *IEEE Journal on Selected Areas in Communication*, vol. SAC-3, no. 2, March, 1985, p. 259.

Taylor, H. F., Giallorenzi, T. G., and Siegel, G. H., Jr., "Fiber Optic Sensors," *First European Conference on Integrated Optics,* September, 1981, pp. 99-100.

Teich, M., "Avalanche Multiplication in Photodiodes," *NSF Workshop and Grantee-User Meeting on Optical Communications*, Ithica, New York, June 26, 1985.

Tsai, C. S., "Guided Wave Acoustooptic Bragg Modulators for Wide-Band Integrated Optic Communication and Signal Processing," *IEEE Transactions on Circuits and Systems,* vol. CAS-26, no. 12, December, 1979, pp. 1072-98.

Tsang, W. T. and Logan, R., *IEEE Journal on Quantum Electronics,* vol. QE-19, November, 1983.

Tsang, W. T. and Olson, N. A., "High-Speed Direct Single-Mode Modulation," *Applied Physics Letters,* vol. 42, 1983, p. 650.

Tsang, W. T., Olsson, N. A. and Logan, R. A., "Optoelectronic Logic Operations by C3 Lasers," *IEEE Journal on Quantum Electronics*, vol. QE-19, November, 1983.

Ubis, S., "Southern Bell Trial Sends Fibers to the Home," *Telephony*, September 8, 1986, p. 25.

Ueno, Y. and Nagura, R., "An Optical Communication System Using Envelope Modulation, *IEEE Transactions on Communications,*, vol-COM20, no. 4, August, 1972.

Van Uitert, L. G. and Wemple, S. H., "Zinc Chloride Glass: A Potential Ultralow-loss Optical Fiber Material," *Applied Physics Letters*, vol. 33, no. 1, 1978, pp. 57-59.

Van Uitert, L. G., Bruce, A. J., Grodkiewicz. W. H., and Wood, D. L., "Minimum Loss Projections for Oxide and Halide Glasses." *Third International Symposium on Halide Glasses*, Renne France, June, 1985.

Warr, M., "Fiber to the Home in 'City of the Future'," *Telephony*, September 21, 1987, p. 13.

Wilson, C., "Southwestern Bell, AT&T Take Fiber to the Home In First Trial," *Telephony*, vol. 213, no. 9, August 31, 1987, p. 10.

Wolfe, J., "Three Telcos Map Fiber Net Plans That May Pose Near-Term Challenge," *Cablevision*, May 18, 1987, pp. 50-56.

Wood, R. and Moore, D., "From Fibrevision to the MultiStar Wideband Network," *Proceedings of SPIE,* vol. 403, Paris, May, 1983, pp. 64-70.

Worthington, P., "Cable Design for Optical Submarine Systems," *IEEE Journal on Selected Areas in Communications, Undersea Lightwave Communications*, November 6, 1984, pp. 873-78.

Wyatt, R. et al., "140 Mbit/s Optical FSK Fibre Heterodyne Experiment at 1.54 Microns," *Electronics Letters*, vol. 20, no. 22, October, 1984, pp. 912-913.

Yamatsu, Y. and Ozeki, T., *Technical Digest of 1977 International Conference on Integrated Optics and Optical Fiber Communications*, p. 371.

Zel'dovich, B. Y., Pilipetsky, N. F., and Shkunov, V. V., *Principles of Phase Conjugation*, Springer-Verlag, 1985.

INDEX